# FIRE PREVENTION ORGANIZATION AND MANAGEMENT

# FIRE PREVENTION ORGANIZATION AND MANAGEMENT

## James Crawford

**Pearson**

Boston   Columbus   Indianapolis   New York   San Francisco   Upper Saddle River
Amsterdam   Cape Town   Dubai   London   Madrid   Milan   Munich   Paris   Montreal   Toronto
Delhi   Mexico City   Sao Paulo   Sydney   Hong Kong   Seoul   Singapore   Taipei   Tokyo

**Publisher:** Julie Levin Alexander
**Senior Acquisitions Editor:** Stephen Smith
**Associate Editor:** Monica Moosang
**Editorial Assistant:** Samantha Sheehan
**Director of Marketing:** Dave Gesell
**Executive Marketing Manager:** Katrin Beacom
**Marketing Specialist:** Michael Sirinides
**Managing Production Editor:** Patrick Walsh
**Production Editor:** Julie Boddorf
**Project Manager:** Susan Hannahs
**Senior Art Director:** Jayne Conte
**Cover Designer:** Bruce Kenselaar
**Cover Art:** Thinkstock and Hughes Associates, Inc.
**Full-Service Project Management:** Sudip Sinha, Aptara®, Inc.
**Composition:** Aptara®, Inc.
**Printer/Binder:** R.R. Donnelley/Willard
**Cover Printer:** Lehigh/Phoenix
**Text Font:** Sabon

Credits and acknowledgments borrowed from other sources and reproduced, with permission, in this textbook appear on appropriate page within text

**Library of Congress Cataloging-in-Publication Data**

Crawford, James,
  Fire prevention organization and management / James Crawford.—1st ed.
    p. cm.
  Includes bibliographical references and index.
  ISBN-13: 978-0-13-508784-8
  ISBN-10: 0-13-508784-8
  1. Schools—Fires and fire prevention.   2. School crisis management.   3. Assistance in emergencies.
  4. Schools—Safety measures.   I. Title.
  LB2866.5.C73 2010
  363.37'9—dc22                                                     2010019281

10 9 8 7 6 5 4 3 2 1

www.pearsonhighered.com

ISBN 10:      0-13-508784-8
ISBN 13: 978-0-13-508784-8

# DEDICATION

To the memory of Katherine Moran Crawford, and to Karen, Ryan, Emily, Rajiv, and Elishuba—my immediate family—for their untiring support of my work.

# CONTENTS

## Chapter 6  Research in Fire Prevention  108

## Chapter 7  Planning for Life Safety and Fire Prevention  122

I firmly believe what my friend Sharif Abdullah once said—that writing for any-one is at once an act of arrogance and humility. Arrogance because it is presumptuous to think we have something important to say. Humility because we recognize that we are just doing our part to improve things, and hope that our efforts are accepted for that purpose. I am very clear on my own role in trying to improve fire and life safety efforts and have humbled myself enough over the years to recognize the truth in this dual view.

I have often said that as a fire marshal in a local jurisdiction for more than 15 years, and with nearly 30 years of experience in the field of fire prevention, I'm just beginning to understand the job. I am constantly reminded that there are people in the field who know more about many aspects of fire prevention than I do. Still, it is important, I think, for all of us to do our part to further the understanding and professionalism of the field. This is my attempt to do so by looking at fire safety and injury control efforts from a comprehensive viewpoint—mostly from a fire department perspective.

There is more detail to be found for each of what I affectionately call the "food groups" of comprehensive prevention programs. Plan review, code enforcement, public education, and fire investigation efforts are all specialized fields in and of themselves. Understanding each is a major undertaking.

My hope is that this attempt to provide an overview will help those who aspire to leadership of these efforts to obtain an understanding of each, and of how they need to work together.

Few things offend me more than someone who continues to view fire prevention bureaus (as they are often called) as a dumping ground for the "sick, lame, or lazy" members of our fire service community. There is a great deal to understand, and not all of it will be found in this book, or any one book. But I hope it is a start, and that readers are engaged enough to understand the complexities of the field—and to pursue even greater understanding as the field advances.

PEARSON
## myfirekit™

As an added bonus, Fire Prevention Organization and Management features a myfirekit, which provides a one-stop shop for online chapter support materials and resources. You can prepare for class and exams with multiple-choice and matching questions, weblinks, study aids, and more! To access myfirekit for this text, please visit **www.bradybooks.com** and click on mybradykit.

# ACKNOWLEDGMENTS

I would like to thank and acknowledge the following reviewers for their insightful recommendations that helped improve this book:

John P. Alexander
*Adjunct Instructor, Connecticut Fire Academy*
*Captain, Hazardville Fire Department*
*Enfield, CT*

Cheryl Edwards
*Fire Safety Inspector/Investigator*
*Lakeland Fire Department*
*Lakeland, FL*

Kevin L. Hammons
*President, Revelations, Inc.*
*Public Safety Services & Solutions*
*Franktown, CO*

Jeffrey L. Huber, M.S., EFO
*Professor of Fire Science*
*Lansing Community College*
*Lansing, MI*

Ernie Misewicz
*Assistant Fire Chief*
*Fairbanks Fire Department*
*Fairbanks, AK*

Chief Terry Peeler
*City of Livingston*
*Alabama State Fire College*
*Livingston, AL*

David K. Walsh
*Program Chair*
*Fire Science Program*
*Dutchess Community College*
*Poughkeepsie, NY*

Michael J. Ward, MIFireE
*Assistant Professor*
*The George Washington University*
*Washington, D.C.*

Gerald E. Wheeler
*NYS Fire Academy (Ret.)*
*Adjunct Instructor*
*Corning Community College*
*Horseheads, NY*

Phil DiNenno
*President, Hughes and Associates, Inc.*

Gary Tokle
*National Fire Protection Association (retired)*

Ozzie Mirkhah
*Las Vegas Fire Department*

There are so many other people to thank for this book that it is impossible to list them all. My wife Karen had to suffer through many nights and weekends of my working on this project, and for her support I'm grateful beyond words. I would be remiss if I did not thank my former Fire Chief, Don Bivins of Vancouver Fire Department in Washington state, for his support of my activities and for his understanding that each of us must try to contribute to the greater good, and that this is part of why we are placed on this earth. Second, to the fine people in the Fire Marshal's Office of VFD—some of the finest and brightest people I have ever had the pleasure of working with. Their support over the years I've been in Vancouver have made me look good—and taught me more than I'll ever be able to describe, or to properly thank them for. John Gentry, Heidi Scarpelli and David Smith all supported me daily and worked harder than anyone else to make Vancouver's fire prevention efforts world class.

This book and its approach were inspired by the Fire Prevention, Organization and Management course at the National Fire Academy, so a great deal of thanks goes to the team who designed that course, especially Ed Kaplan and the other people who developed the original course material that I used and expanded upon. Their work and the efforts of the Fire and Emergency Services Higher Education prevention team was groundbreaking in my opinion.

I'd also like to thank Senior Acquisitions Editor Stephen Smith and Associate Editor Monica Moosang of Pearson Education, and Kristin Landon, an independent copy editor, for their hard work in getting this project off the ground and refined.

Special thanks to the many friends who have inspired me over the years and/or provided material for this book.

Jim Marshall, *Ph.D. Professor (retired). Portland State University, Portland, Oregon*

Meri-K Appy, *President of the Home Safety Council*

Sandra Facinoli, *Branch Chief of Fire Prevention at the U.S. Fire Administration*

Steve Sawyer, *Executive Director of the International Fire Marshals Association*

Ed Kaplan, *Fire Program Manager of the National Fire Academy*

Dan Uthe, *Fire Marshal of the Tucson Fire Department*

Woody Stratton, *Fire Program Specialist of the National Fire Academy*

Azarang (Ozzie) Mirkhah, *Fire Protection Engineer of the Las Vegas Fire Department*

Wayne Powell, *Fire/Life Safety Specialist of Marriott International*

Ron Coleman, *retired State Fire Marshal of California*

Gerri Penny, *Public Education Coordinator of the West Palm Beach Fire Department*

LaRon Tolley, *Fire Program Manager for the Division of Extended Studies of Western Oregon University*

Peg Carson, *of Carson-Associates, Inc.*

Vicki Hamilton Wade, *Fire Program Specialist of the Department of Homeland Security*

Ken Lauziere, *Special Consultant of the Cabezon Group*

Earl Diment, *of Pioneering Technologies, Inc.*

Rob Ware, *(retired) of Portland Fire & Rescue in Oregon*

Don Porth, *of Portland Fire & Rescue in Oregon*

Angela Mickelide, *Ph.D., Director of Education and Outreach of the Home Safety Council*

Steve Zaccard, *Fire Marshal for the St. Paul, Minnesota, Fire Department*

Paula Peterson, *Public Education Coordinator for the St. Paul, Minnesota, Fire Department*

Phil Schaenman, *President of TriData Corporation—a division of System Planning Corporation*

Dr. John Hall, *Vice President of Research at the National Fire Protection Association*

This book would not be possible except for the work that has preceded it—and the help and guidance of all those who contributed. I am deeply grateful.

# ABOUT THE AUTHOR

**Jim Crawford** is the retired Deputy Chief and Fire Marshal for Vancouver, Washington, where his responsibilities included business planning, code enforcement, plan review, fire investigation, and public education activities. Mr. Crawford served as the Fire Marshal in Portland, Oregon, before moving to Vancouver. With more than 35 years of experience in the fire service, with the last 18 in senior-level management, Mr. Crawford has been an active participant at the national level with the U.S. Fire Administration, the International Fire Marshals Association (past president), the International Association of Fire Chiefs, and the National Fire Protection Association . He serves as the current chair for the technical committee on professional qualifications of fire marshals (NFPA 1037). He is the author of *Fire Prevention: A Comprehensive Approach,* published by Pearson Education and is a regular columnist with *Fire Rescue* magazine. He has served as adjunct faculty at the National Fire Academy and is the project director for Vision 20/20 (www.strategicfire.org), a national strategic plan for improving fire prevention efforts in the United States.

Mr. Crawford holds an A.S. in Fire Science from Clackamas Community College (OR), and a B.S. in Management and Communications from Concordia University.

The following grid outlines Fire Prevention, Organization and Management course requirements and the chapters in which specific content can be located within this text:

| Course Requirements | 1 | 2 | 3 | 4 | 5 | 6 | 7 | 8 | 9 | 10 | 11 | 12 | 13 | 14 |
|---|---|---|---|---|---|---|---|---|---|---|---|---|---|---|
| Describe aspects of risk reduction education and overall community risk reduction. | X | X | | | | | X | | | | | | | |
| Explain the fundamental aspects of codes and standards, and the inspection and plan review process. | X | | X | X | | | | | | | | | | |
| Describe the fire investigation process and discuss fire prevention research. | X | | | | X | | | | | | | | | |
| Discuss historical and social influences and describe the master planning process. | | | | | | | | X | X | | | | | |
| Describe economic and governmental influences on fire prevention. | | | | | | | | | | X | X | | | |
| Explain the effects of departmental influences on fire prevention programs and activities. | | | | | | | | | | | X | | | |
| Discuss strategies for fire prevention. | X | X | X | X | X | X | X | X | X | X | X | X | X | X |

his wildfire threatened
ancouver neighborhood...

**Are You Prepared?**

# 1 Concepts of Fire Prevention

## KEY TERMS

Administrative duties of the fire
marshal *p. 22*

Coalition development *p. 4*

Code enforcement *p. 8*

Investigation *p. 4*

Juvenile firesetters *p. 15*

Plan review *p. 4*

Public education *p. 4*

## OBJECTIVES

After reading this chapter you will be able to:

- Identify the elements of a comprehensive prevention program, including plan review, code enforcement, public education, coalition development, and investigation.
- Identify the elements of juvenile firesetting intervention programs.
- Understand the administrative nature of the fire marshal's job duties.
- Understand how fire research sets a foundation for fire prevention efforts.

PEARSON
## myfirekit™

For additional review and practice tests, visit **www.bradybooks.com** and click on MyBradyKit to access book-specific resources for this text!

# An Overview of the Comprehensive Approach to Fire Prevention

The *Funk & Wagnalls Standard Desk Dictionary* defines prevention as follows: "to keep from happening, as by previous measures or preparation; preclude; thwart." Most people understand that preventing emergencies is the most cost-effective way to deal with them. However, few in our modern industrialized society place much emphasis on prevention. Our attention is usually focused elsewhere, until a catastrophic event has already occurred. Then attempts at prevention are too late. Yet the tragic lessons of our past compel us to take at least some measure of caution in our daily lives. Most children are taught that knives are sharp and can cause harm. But fire and other catastrophic emergencies are a little more complicated than a sharp knife, and sometimes the hazards are less visible to the untrained eye.

Throughout human history, fire has been feared for its potential to cause great destruction. Efforts to control and even prevent fires have been documented from the early days of the Roman Empire. History has abundant lessons to teach about the tragedy of fire once it is out of control. The great fires in London and Chicago are well-known events, even to those outside the fire protection field. But closer to our modern age are the significant fires of the twentieth century that have helped shape more modern fire prevention efforts.

For example, on March 25, 1911, a tragic fire occurred in the Asch building in New York City. Commonly referred to as the Triangle Shirtwaist fire, it resulted in the deaths of 146 young women who worked in the garment factory. Many died jumping out of windows trying to avoid being killed by the fire. Most attribute that fire and ensuing efforts to promote more effective prevention efforts are principal factors in the establishment of the *Life Safety Code* produced by the National Fire Protection Association (NFPA).

Other tragic examples include the New London, Texas, gas explosion of 1937. In this incident, a large accumulation of gas ignited, and the resulting explosion killed 294 schoolchildren and teachers. The Cocoanut Grove nightclub fire in Boston, Massachusetts, occurred in 1942, killing 492 people in an overcrowded club with limited exits and combustible walls and decorations. In Hartford, Connecticut, the Ringling Brothers and Barnum and Bailey circus tent fire of 1944 killed 168 people.

More recently, the terrorist attacks on the World Trade Center towers in New York (September 11, 2001) provided another tragic example where thousands died in building collapses caused by fires that proved too intense for the built-in fire protection features. This provided us many lessons about adequate exiting and protection of steel structural supports. And the Station nightclub fire in Rhode Island on February 20, 2003, in which 100 people died, taught us lessons about the importance of effective fire inspections, the risks of using combustible and flammable materials for decorative purposes or acoustic features, and the fact that many will try to exit in an emergency from the same point at which they entered a structure, even if other exits are available elsewhere.

These examples of multiple-death fires, and many similar tragedies, have led to the development of a more modern approach to prevention. In one sense, each tragedy points out that we are still learning from disasters of the past, and perhaps

The Station nightclub fire in 2003 was another major life-loss fire that helped shape modern fire prevention codes. (*Courtesy of NIST*)

not learning enough. But over time, the definition of prevention in the context of fire safety has changed, as well as its ramifications for the fire service in its evolution as a public safety agent.

Traditionally, prevention efforts in the fire service have been somewhat limited. Usually, the emphasis is placed on fire code enforcement inspections and fire investigations, with some activity devoted to educating the general public about the danger of fire and how it can be prevented. But the term *prevention* has been expanded by the fire service to include those proactive measures that actually prevent a fire or those that prevent its spread once a fire does occur. However, new challenges have caused another evolution in thought about what *prevention* means to the fire service, ironically fueled by some of the changes brought about by successful fire prevention efforts.

Today, fire departments in North America handle fewer fires and are now called on to deal with a wide variety of other emergency incidents. Consequently, the prevention efforts in most fire departments are really efforts to control losses of human lives and property damage from a variety of causes. Most calls handled by a modern fire department (as many as 70 percent in some cases) are medical emergencies. In addition, fire departments respond to a variety of hazardous material emergencies that threaten public health or the environment (or both). A modern fire department's calls for emergency assistance include drowning incidents, auto accidents, trench rescues, terrorist actions, and numerous other emergencies. As a result of this trend, many departments are increasing their prevention efforts to deal with these community needs. Departments are responding to a public need for prevention efforts that deal with all types of injuries, not just those caused by fire.

## ENGINEERING, ENFORCEMENT, EDUCATION, EVALUATION (AND MORE)

The building blocks for prevention efforts are essentially the same for each type of emergency. Efforts to control losses are generally categorized as *engineering*, *enforcement*, or *educational* mitigation measures. They constitute the basics of programs designed to reduce risk or actual loss. In practice, these three concepts are generally divided into specific functional areas that many fire departments use to prevent losses. For example, reviewing plans for new construction would constitute a use of *engineering* principles in construction or fixed fire protection to prevent fires or fire spread. This is accomplished during a construction **plan review** by making sure building features meet code requirements or that alternatives such as fixed fire protection are identified and included in the design before construction begins.

Using the *enforcement* function as a prevention tool assumes that there is a need for laws and rules that govern construction and behavior that will lead to reduced risk or loss. This naturally calls for the development of those laws and rules and an ability to enforce compliance with them. However, many fire departments are finding that their most significant problems occur in one- or two-family dwellings where efforts to design or regulate safety clash with a culture that values personal privacy and rights above all else. Consequently, **public education** has become a primary tool in ensuring safe behaviors that also reduce risk and loss from a variety of emergencies. In this context, *education* refers to a voluntary method for reaching and teaching people how to be more fire safe.

Underlying these loss-control efforts is an assumption that active **investigation** of the *causes* of emergencies will lead to better prevention efforts. As a result, most fire departments also pursue some type of fire investigation activity as part of their prevention efforts. Many are also actively involved in the prevention of arson; therefore, investigation programs usually contain elements that address the criminal investigation and prosecution of arson.

These basic elements combine to create a more comprehensive approach to prevention than is traditionally found in many fire departments. Comprehensive approaches to fire prevention reflect the trend toward multifaceted ways of preventing fires from occurring, or limiting the damage once they do occur. A *comprehensive approach* generally refers to combined efforts for plan review, code enforcement, public education, investigations, all-injury prevention programs, and **coalition development** to increase prevention resources, as well as efforts to evaluate them.

The comprehensive scope of effort is beyond the reach of some departments, but many are finding unique ways to accomplish the same objectives. For example, some departments are finding that involving the community in solving their own problems is the only way to control losses in an effective fashion. Establishing coalitions with business and community groups has become a more common practice and has proven to be a proactive way for a local community to take ownership of solving its own emergency problems.

**plan review**
▪ checking construction plans for compliance with applicable codes and conducting acceptance inspections in the field to ensure that construction practices match what is shown on the plans

**public education**
▪ results-oriented programs that raise the cognitive awareness levels and ultimately change behaviors of targeted audiences

**investigation**
▪ activity by which cause and origin is determined; generally refers to fires and explosions

**coalition development**
▪ the process of bringing partners outside the fire service together to support fire prevention programs and goals

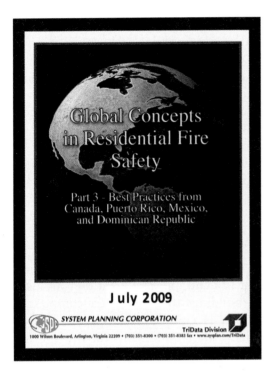

TriData and the Centers for Disease Control have conducted several international studies demonstrating which prevention programs produce measurable results. (*Courtesy of TriData Corporation, a Division of System Planning Corporation of Virginia*)

There are factors that hinder efforts to establish comprehensive prevention programs. According to the National Fire Protection Association, the United States and Canada are still among the worst of industrialized nations at controlling losses. One private consulting firm based in Virginia (TriData Corporation, a Division of System Planning Corporation of Virginia) has produced several reports[1] that examine why European and Asian nations tend to place more emphasis on prevention. The long history of fire in European and Asian countries has resulted in a strong emphasis on safety. The United States and Canada (and some other industrialized nations) still emphasize technological solutions to most problems and have a cultural bias that nature (and fire) can be controlled through external means. This belief in technology, combined with a consumer culture that uses materials and discards them, creates an apathy in our populations that relegates prevention to a lower priority.

At the same time, exponential growth in science and technology increases the pressure on fire departments to stay abreast of the hazards that these new technologies represent to any given community. The use of toxic chemicals and products increases the potential for hazardous material emergencies that can threaten large portions of a community. In addition, demographic changes lead to cultural and behavioral differences that can increase the demands on a fire department's resources. There are many examples of immigrants who have misused heating or cooking appliances that are unfamiliar to them. Prevention efforts for these populations have created a demand for multicultural and multilingual programs that few departments are prepared to produce.

Generally speaking, a community that is more complex (both technologically and culturally) will make the establishment of comprehensive prevention efforts more challenging. But whether small or large, communities must use these basic loss-control tools and meet these challenges to have an effective comprehensive prevention program.

It is important for fire service leaders and other decision makers to understand that modern prevention programs have taken a new shape and require a great deal of concentrated effort if they are to be effective. A comprehensive prevention program is designed to manage risk and loss for a variety of emergency incidents through the use of basic tools common in loss-control efforts.

This chapter is devoted to providing an overview of these basic prevention components and describes some of the issues that must be addressed in order to develop and maintain a comprehensive prevention program. The remainder of the book will be devoted to an examination of each of these topics in greater detail, but an overview is appropriate to gain an understanding of how they are interrelated.

## THE PLAN REVIEW PROCESS

Many fire departments do not participate directly in the review of construction plans that help to ensure a certain degree of fire and life safety. Yet they are usually responsible for inspecting these same buildings to ensure their compliance with applicable fire codes after construction is completed. Inspections often require a working knowledge of the building code requirements for fire and life safety, and an inspection ensures that these features of construction are functional. This overlap between the fire and building codes is generally recognized by the people who develop the model codes, and considerable effort is made in correlating them for the sake of consistency and public safety. Consequently, the plan review process involves elements of engineering safety into the construction of buildings and enforcing the laws and ordinances that keep them in operation.

The concept of reviewing plans and inspecting buildings for code compliance *before* they are issued certificates of occupancy is core to engineering safety into modern buildings. A certificate of occupancy is usually the last step in the construction process, held until the building meets all applicable codes and standards.

The combination of these prevention efforts must be coordinated at the local level. The fire service should have an active interest, if not direct participation, in the plan review process. If buildings are not watched closely during the construction phase, then even modern codes driven by all the previous fire disasters may have little impact—simply because they may not be included in the construction and there is no one to point out when they are missing.

There are many fire and life safety issues examined during the building plan review process of interest to the fire service. These issues include emergency vehicle access requirements, water supplies for firefighting, fixed fire protection features such as fire sprinklers and alarms, and proper emergency exits. The process of reviewing plans for new construction provides the fire department with an opportunity to prevent fires through proper construction methods or to mitigate the effects if one does occur.

There is also an interrelationship between the construction and building features of concern to the fire service and those of other departments. Those responsible for transportation, environmental, or planning codes also interact with fire service goals. For example, when urban areas extend into the wildland, the requirements for emergency-vehicle access become more stringent. People escaping in automobiles from a wildland fire often encounter emergency vehicles on their way to protect their homes, so wider streets are necessary to accommodate both residential and emergency traffic. In addition, issues of vegetation control, roofing construction, and water supplies are factors in these situations. These issues are often encountered in local planning and environmental codes that deal with where development may occur and how it may affect the natural environment. Transportation codes exist to control traffic and provide for safety by narrowing streets and placing traffic control devices that can slow emergency responders. Environmental codes want to reduce nonpermeable surfaces to aid water quality and storm water runoff, so they generally support narrow streets. Abundant vegetation helps reduce water pollution and improves air quality; therefore, those responsible for environmental codes may resist efforts to cut or control vegetation for fire safety purposes. In other words, the interests of other parties within the community may conflict with the needs and concerns of the fire department.

Another code that ties closely with the fire code is the mechanical code. It controls heating and ventilation issues during construction. Smoke patterns during a fire are affected by the ventilation systems, which can either help or hinder smoke management during a fire. Improperly installed or maintained heating systems are often the cause of fires.

The electrical code is designed for public safety and helps limit the number of fires caused by electrical malfunctions. Proper grounding procedures have been in place for many years to prevent electrical problems that can cause fires, and recent technological developments such as arc fault circuit interrupters (devices that cut current when electrical arcs are detected) may also help. But these safety features are driven by a code that is not usually regulated by fire departments.

To briefly summarize, the construction process is a critical part of a comprehensive prevention effort. It provides the opportunity for the fire service to ensure that proper controls are engineered into the structure from the beginning. The lessons

An inspector reviewing plans for new construction for compliance with appropriate building and fire codes. (*Courtesy of Gary Styers, Mooresville Fire Department*)

learned from past disasters have influenced the development of model codes that deal with safety, and the fire service must be aware of the interrelationship between these various codes and their competing interests within the community. Consequently, the fire service should pay close attention to and participate in the plan review process, and be knowledgeable about the related codes that can affect the nature and size of a fire problem for any community.

## CODE ENFORCEMENT AND DEVELOPMENT

Another major part of a comprehensive prevention program is that which deals with *enforcement* of the codes or laws relating to fire and life safety. Laws or codes put in place with the best intentions will accomplish nothing if they are not followed. The nature of the **code enforcement** process usually involves a measure of educating the business community about the requirements in the fire code, most of which are usually unknown to them. In this fashion, the *enforcement* and *educational* elements of a comprehensive prevention program also overlap. However, not everyone sees the same need for fire and life safety measures, and so at times the enforcement of the codes is necessary.

**code enforcement**
■ the process in which local jurisdictions have the authority to order compliance with fire and/or building code requirements

There are several fundamental issues about code enforcement that must be understood by fire officials and other local decision makers. Among them are the authority to adopt and administer codes, the management of an appeals process, the development of codes and laws, the developing trend of performance-based codes, and the globalization of codes and standards.

The authority to adopt and administer codes is usually derived from the state. Model codes are produced by a variety of organizations that can be used as the basis for local code adoption, but local jurisdictions must have a clear plan for legally adopting their own version and establishing the limits of their authority. The administration of the code enforcement process will not generally preclude the authority of the judicial system; therefore, local decision makers should obtain proper legal counsel to help establish a code adoption and administration process that will withstand the inevitable legal challenges by those who question the authority of the jurisdiction for code enforcement activities. The municipality or county involved in adopting fire and life safety codes usually obtains their own legal representative, either a full-time employee or a lawyer on retainer.

Once fire and life safety codes are legally adopted, local decision makers and fire officials must give thought to the proper administration of their enforcement. It is virtually impossible to ensure that every occupancy that falls under the jurisdiction of the codes is in complete compliance with them. As a result, the code enforcement process for any fire department usually involves an inventory and prioritization of the properties to be inspected. Once that is accomplished, it is important to develop and maintain records of inspections and create a database that will track the identification and abatement of hazards or violations of the codes. This information will be critical if a fire or other related loss does occur in an occupancy that has been inspected, because that fact will mean that the local jurisdiction might assume a portion of the liability for the damages. Consequently, keeping track of violations and making sure they are abated is a necessary protection against a claim of negligent inspection practices by someone wishing to spread the economic loss produced by a fire or other emergency.

An inspector conducting field inspections of existing occupancies for compliance with fire code regulations. (*Courtesy of Virginia Department of Fire Programs—Virginia Fire Marshal Academy*)

The inspection process may include a variety of schedules and approaches to staffing that are discussed in more detail in Chapter 3. Use of designated staff, engine company personnel, and consideration of when inspection cycles or the time of day they are conducted are all factors to consider when managing the code enforcement process.

Administering the code enforcement process also means selecting and training personnel to do the job. Untrained or unprofessional inspectors can miss hazards and increase the risk of liability for their department. In larger jurisdictions with many inspectors, consistency of inspections is always a concern. Because the codes are complex, more experienced inspectors usually identify more hazards. This can lead to complaints from occupants who state that previous inspections failed to identify a hazard that is now being pointed out. Proper selection and training of the inspectors can help keep all the inspection personnel performing at an appropriate level and in compliance with their directives, as well as increase consistency.

In other words, the proper administration of a code enforcement process is complex and requires specially trained personnel to successfully manage it.

Each of the model code organizations offer some type of training and certification for fire inspectors that will also ensure the quality of inspections. Some fire departments produce checklists for their inspectors to use in an attempt to ensure consistent inspections among different personnel. Some also use preinspection letters to notify businesses about common hazards so that the businesses can become proactive partners in the hazard abatement process. Adequate training and use of checklists and preinspection letters may help to ensure that inspectors look for hazards in a consistent fashion that reduces the chances that some hazards may be missed. Certification of those responsible for conducting code enforcement activities is critical and should follow nationally recognized professional qualification standards. For code enforcement and plan reviewers, NFPA 1031 has appropriate information about critical job performance responsibilities.

Managing appeals to the code is also part of the administration process. Many parts of the codes are subject to interpretation. In addition, they usually allow alternate materials and methods that will produce essentially the same level of safety. Consequently, many opportunities arise to challenge the code interpretation or to argue that some feature not prescribed in the code will actually produce the same level of protection.

To handle challenges to the code, a board of appeals is usually established to act as the final authority (before court) for fire code administration. The board of appeals can be established at the state or local level and is usually composed of architects, engineers, and design professionals from the community who have a working knowledge of the code but are not directly connected with the fire service. Their purpose is to act as the final decision maker when interpretation or alternate methods are at issue.

Many fire departments also establish an internal review board that can handle the task of resolving many of these problems before they reach the formal board of appeals. Managing appeals means processing them in a timely fashion and preparing board members and code management personnel to act within the scope of their authority. The appeals board may also be responsible for determining which alternative solutions to prescriptive codes provide the same level of safety, but more is said about that topic and performance codes later in this book.

Adopting codes, laws, and policies for fire and life safety implies that they must be kept updated and in concert with modern safety practices. To do so, those adopting codes have an obligation to participate in the development of the model codes and standards. There is no better advocate for quality codes than the people who administer them daily and understand their effect on the community. Too often local decision makers misunderstand the need for code administrators to participate in the code development process and do not provide the resources necessary for them to do so. Knowledge of past disasters combined with the experience of those in the field who are responsible for code enforcement is a compelling argument that local involvement in the development of quality codes is not a luxury, but a requirement.

One of the more recent issues in model code development is the movement toward performance-based codes. These are codes that are written with a specific safety objective in mind, rather than the alternate method of prescribing specific safety features. In the United States, these performance code issues are still being tested, and the computer models used to determine an equivalent level of safety are still in their infancy. For years the fire codes have traditionally allowed alternate materials and methods to achieve similar levels of safety, but performance codes attempt to do so on a much larger scale. A variety of model code development organizations, including the National Fire Protection Association, the International Code Council, and the Society of Fire Protection Engineers, are actively working on the refinement of performance-based codes.

As a result of this trend, local fire officials must be prepared to elevate the importance of performance codes in their code enforcement programs. This will usually mean providing a higher level of training and certification for code enforcement personnel. Many fire departments are hiring fire protection engineers to conduct their plan reviews and assist with their code enforcement activities.

The level of engineering and scientific expertise needed to assess the performance standards under this type of code enforcement system will continue to grow.

Another emerging trend in fire and life safety protection is the internationalization of codes. Trade agreements between nations open up economic opportunities for international companies. These companies desire consistency of codes across international boundaries. Because of this trend, most code and standards development organizations are producing model documents that can be used in more than one nation. As the demand for these codes and standards increases, the people who participate in their adoption and development also increases. Specific countries with their own economies have opinions about how codes should be developed, based on their own experience and ability to afford safety features. Some have very different views about how many safety features should be provided and what constitutes an acceptable level of risk. Many developing countries cannot afford the same level of fire protection offered people in more industrialized nations, but their votes may count equally in the code development process.

Those with a stake in code development are now being faced with other points of view from the broader international arena. Fire officials and other local decision makers must be aware of other viewpoints and meet them with active participation of their own. The development of quality codes relies on sound science, engineering principles, and accurate data to ensure proper safety procedures within this new global arena.

To briefly summarize, code enforcement is a necessary part of any comprehensive prevention effort. It often overlaps with the engineering and educational efforts, which are also designed to improve public fire and life safety. Fire officials and local decision makers must understand the limits and the legal foundation of their code enforcement activities. They must understand the elements of an effective administrative process, which includes an inventory of properties to be inspected, a database and record keeping system, selection and training of personnel, and management of the appeals process.

## PUBLIC EDUCATION AND RISK REDUCTION

Most fires, fire deaths, and other injuries occur in one- and two-family dwellings where code enforcement abilities are severely restricted. Even individual apartment units enjoy a good deal of protection from code enforcement activities because of a basic precept of personal privacy. But most fire deaths occur in residential properties; therefore, public fire and life safety education is one of the more important strategies of a comprehensive prevention program. Many fire departments are beginning to understand the value of public education and what it can produce for their total protection and prevention efforts. They are increasing their resources for this vital function and producing positive results. The basic purpose of public education efforts is to increase knowledge and change behaviors to reduce risks and losses.

There are two broad approaches to a modern public education program. The first is general education, such as programs designed to provide broad safety messages within the school system. The logic behind this approach to fire and life safety is that reaching children while they are young will produce results that last a lifetime. The second approach is to be more targeted to specific audiences and specific messages, such as those that go directly to portions of the public, reaching them through a variety of methods and media. These programs have the same basic tenet in mind: to increase public knowledge and change behavior to improve public safety; however, they are more focused than general public education programs.

## Education in the Schools

Schools are generally recognized as a critical part of any public education program. Many local decision makers have begun to understand that schools are under increasing pressure to improve their performance while reducing costs, and fire safety may not receive the priority it needs because of those concerns. Whatever the constraints, public fire and life safety education must go beyond simple presentations about what a firefighter looks like in protective gear. Showing schoolchildren what firefighters do has good public relations value, but that does not teach children how to be safe. Some programs include well-designed educational activities that use firefighters or public education personnel to teach children directly. Others utilize teacher curricula and well-thought-out educational strategies to reach children and change their behavior.

As a result of the pressure on schools to perform more with less and the sophistication level of public fire and life safety education increases, many fire departments are also changing their public education efforts in schools to include other injuries in an attempt to maximize their ability to impact children's safety behaviors with limited resources. This all-injury prevention approach to safety curriculums is often necessary in order to reach students in an efficient manner.

One example of a all-injury safety curriculum is the *Risk Watch*® program developed by the National Fire Protection Association.[2] It covers eight basic injury-control topics, including fire, water, firearm, and bicycle safety. It was

An elementary school fire safety presentation. (*Courtesy of George Armstrong/FEMA*)

developed around the concept of using a *coalition of organizations* to promote safety for schoolchildren. It presumes that the fire service, police, public health, and other agencies with an interest in public safety must cooperate to maximize their impact in schools already pressed for time and resources. The *Risk Watch* curriculum is no longer a principal product of NFPA and has limited support for distribution. But it still serves as a model example of an all-injury control safety curriculum. Other agencies, such as the Red Cross and some local jurisdictions across the nation, have developed similar school safety curricula. The most successful are those that can demonstrate results by documenting increases in knowledge and safety behavior changes.

This more modern approach to fire and life safety education in the schools involves students performing activities that lead them to make proper choices about safety. Research indicates that students who are actively involved in decision making are more prone to remember educational messages and change safety behaviors. The whole concept of using a curriculum for educational purposes is based on a simple principle: teachers are the best resource for providing education in the schools. The fire service can more effectively accomplish its prevention goals by assisting teachers instead of trying to replicate their role in the schools.

## Coalition Development

An important issue for those pursuing effective educational strategies in schools is that there is a clear movement toward preventing all the injuries a modern fire service agency encounters. Consequently, there is a national trend toward the establishment of coalitions between the fire service and other organizations to more effectively achieve the objective of raising the public consciousness and level of concern for safety by doing so *collectively.* Fire departments, police departments, public health agencies, safe-kids coalitions, the Red Cross, and others are beginning to become more effective partners in fire and life safety education with the fire service.

Education anywhere, but especially in schools, must achieve a level of sophistication that truly teaches students proper messages and behaviors. Lessons should include solid educational activities that involve the students in decision-making and learning exercises. Lessons must be well tested and documented to provide evidence that the resources provided actually produce something of value for the schoolchildren and the departments investing in them.

Quality products for school or other educational programs have been tested and the educational value of their material has been proved. There are other safety curriculum products available for use. But any curriculum developed at the local or state level must not only stand up to the same type of testing, it must also be in a form that schools can easily use or it will gather dust in storage.

Few schools would turn down assistance in the form of donated fire and life safety curricula. But for a curriculum to be useful, a fire department must support it and reinforce the messages with presentations that reiterate the safety consciousness children achieve with a well-structured educational program. Special presentations such as clown programs or puppet shows and other age-appropriate educational offerings can reinforce the proper messages for children. And a simple visit from the local firefighters can serve as a positive reward mechanism for teachers to help them motivate children's learning.

### Targeted Public Education Programs

The other major part of a public education program is to create messages for the general public outside the school system. Doing so usually means using one or more media to reach them, and segmenting the audiences we are trying to reach. Presentations at civic events, press releases, or public service announcements through the news media are examples of how many departments attempt to reach the adult population. Because of the lack of resources generally applied to public education programs, departments are usually forced to prioritize and target their efforts.

In the past, many public education programs were chosen for their popularity or because someone had an interest in them. No regard was given to what the local data showed about a community's fire or injury problem. Clown programs and puppet shows became the norm for some fire departments, with no effort made to determine a true target based on available data most fire departments collect. These programs are most appropriate for children, but educating children is only part of a comprehensive public education strategy.

Local jurisdictions need to analyze their loss data and pick the most appropriate public education strategies that will work to control those losses. This analysis usually leads to the conclusion that there are several other target audiences besides schoolchildren. Aside from the young, national statistics[3] show that the other targets are usually the elderly, ethnic minorities, and the poor. In fact, the strongest correlating factor for fire losses is the income level of victims. Those at the lower end of the socioeconomic scale have more fires.

Targeted educational programs are most effective when they are specifically developed for each target audience and target message. They are often combined with other prevention strategies for maximum effect. In any comprehensive prevention program, engineering and code enforcement efforts include educational

An example of a targeted public education mailer for a specific message and audience. (*Courtesy of Vancouver Fire Department, WA; Michael Andrew Dalton, photographer*)

elements. Teaching people how to abate hazards before an inspection is one such example. Educating people about the value of fire sprinklers is another example. Successful programs are designed with the audience in mind, and messages are repeated frequently enough to be absorbed in an environment where competition for consumer attention is great.

Targeted educational efforts extend beyond the typical school setting, using modern marketing venues to reach the population where they live. Radio, television, and newspapers all play a role in this marketing effort. Many fire departments have also had success using grassroots efforts, such as door-to-door canvassing, to deliver their message. In fact, the use of home safety surveys is proving to be one of the most effective prevention efforts available. Recent studies by TriData and the Centers for Disease Control and Prevention have provided evidence of significant drops in fire loss rates for nations actively pursuing this educational strategy.

Specialized displays to help demonstrate the value of fire sprinklers or home fire safety "trailers" have also become popular ways to provide education to certain segments of the community. Many departments still provide specialized presentations to business or civic organizations or participate in large events, such as home and garden shows, to get the fire and life safety message to the public.

## Juvenile Firesetting Issues

Children are a specific audience of note for other reasons. **Juvenile firesetters** are a specific fire problem that requires a multidisciplinary approach, including juvenile justice, mental health, and school partners. Effective education programs can reduce the incidence of child firesetting by teaching children (and parents) that matches and lighters must be kept out of the reach of small children. But aside from education, families of firesetters need to be psychologically screened to determine the level of firesetting activity of their children, and in some cases families may require psychological intervention to prevent recurrence of the firesetting behavior.

In those cases, it is not enough just to educate children to avoid playing with matches or lighters. Emotionally disturbed children act out with fire for a variety of reasons. Until they are identified, and referred to adequate counseling, the firesetting behaviors are likely to continue and may even worsen. Thus, a partnership between fire departments and mental health providers is essential for a juvenile firesetting prevention program.

However they are delivered, public fire and life safety education programs tend to work best when they involve the partner agencies who already have concerns about a particular audience or problem. Having a comprehensive public fire and life safety education program means providing true educational opportunities in a variety of settings. Whether presented in schools, in community meetings, or through marketing outlets, quality education must be age appropriate and in a form that will capture interest. It should raise the level of consciousness for the public about safety. However, educational efforts must ultimately change the behavior of targeted populations to reduce the risk and loss from fire and other injuries.

juvenile firesetters
■ younger children who most often set fires out of a sense of curiosity without an understanding of the consequences, but occasionally do so because of some emotional disturbance

## INVESTIGATION

The final major component of a comprehensive prevention program is the investigation function. It is the underlying foundation that provides the information necessary to guide prevention efforts. Although not specifically part of engineering or educational solutions, fire investigation is the basis on which these prevention strategies are built. For many fire departments, the fire investigation unit is also responsible for arson determination and assists in its prosecution. Consequently, investigation is a direct part of the enforcement strategy for a comprehensive prevention program.

Most fire departments investigate fires to identify the area of origin and determine the probable cause, though the level of sophistication with which they do so varies. If the cause is readily evident, responding fire officers can successfully conclude how a fire originated. If the cause is not so clear, more highly trained investigators are needed.

The basic purpose of investigating fires, therefore, is to help determine what caused the event so that it can be prevented in the future. However, fires are also investigated to determine whether or not the crime of arson has been committed. If a crime is suspected, the nature of the investigation changes, and law enforcement officials become primary partners in the investigation procedure. The police and public prosecutors may take an active interest in the investigation, and interviewing witnesses may become more difficult. Many fire departments turn the investigation over to law enforcement agencies entirely when a crime is suspected. Others train and certify their investigators as police officers in order to help maintain continuity during criminal investigations.

If a crime is not suspected, there are still elements of an investigation that may involve outside interests. For example, most insurance companies routinely hire their own investigators to determine the cause of a fire. They do this in part to protect themselves from civil actions, such as subrogation suits between insurance companies or product manufacturers, over which agency has the responsibility to pay for the damage. The manufacturer of an electrical appliance that caused a fire may be sued by the insurance carrier to recover expenses for the damages of the actual fire and the expenses of the investigation. Under these circumstances, private investigators hired by insurance companies may or may not agree with local authorities about the cause. When a very expensive fire loss occurs, there are usually enough special interests with a stake in the financial loss to make the investigation complex and subject to scrutiny from many outside sources. These circumstances demand a high level of proficiency for fire investigators and a solid presentation of scientific evidence to support claims about cause. This practice is driving many fire departments to increase their proficiency in fire investigation.

Whether a fire occurs through unintentional events or is purposely set, there are some common activities to its investigation. An investigation starts with observations about fire conditions from those who first arrived at the scene. Firefighters first on the scene may observe characteristics of smoke or flame color and other factors that may aid trained investigators to determine a fire's cause. Firefighters must also preserve the scene of the fire as much as possible to ensure that critical physical evidence will not be destroyed. Before determining a fire's cause, investigators must first determine the area of origin for the fire. Examination of burn patterns can lead investigators to the area of origin, even in a badly damaged structure. Once the area of origin is determined, an understanding of how physical items in a structure react during a fire can lead to identification of its probable cause.

A fire investigator digging through debris at a fire scene to begin pinpointing the cause. (*Courtesy of Montgomery County Fire & Rescue, MD*)

Fires are caused by many different events. They are often caused by some electrical malfunction. Carelessly discarded cigarettes are also a principal cause of fires and fire deaths. Many fires also occur in the kitchen because of poor cooking practices, such as leaving cooking food unattended. Whatever the cause, it is always some combination of a heat source coming in contact with something that will burn. In addition, human actions, deliberate or unintentional, are almost always a significant factor in a fire's cause or spread.

Determining the cause of a fire and the relationships among all the factors that contribute to it can require very specialized expertise. Because it is the mission of a fire investigation program to effectively determine cause, those performing the investigation must be well trained to recognize fire burn patterns and apply the science of investigative techniques to the physical characteristics of the many materials inside a fire scene. Training is not enough. Adequate fire investigation efforts also require the equipment necessary to identify causes and aid in prosecution of crimes. For example, many fire investigations require laboratory reviews to corroborate an investigation's conclusion. Many departments are also employing the use of specially trained accelerant-detection dogs to help pinpoint the use of accelerants. This information, when properly validated, can help in the prosecution of arson cases.

Investigators must also be trained to interview witnesses of the fire to help recreate the fire scenario. This is particularly important when arson is suspected.

In such cases, investigators must preserve any evidence and establish a "chain" of its possession so that the evidence can be used in any future court proceedings. Arson investigations require special interview procedures and documentation to provide a conclusion that will withstand legal scrutiny.

During arson investigations, the relationship among investigators, the local police, and prosecutors is critical. A report done for the U.S. Fire Administration by TriData Corporation[4] concluded that because cause determination, arson investigation, and the development of a criminal or civil case occur in sequence, close coordination between fire and police agencies is particularly important. Some jurisdictions are creating task forces of multijurisdictional teams to handle more complex fire cause investigations, particularly where engineering expertise is needed or complex law enforcement issues arise. More is said about investigation in Chapter 5.

The results of a fire's investigation can also contribute to the database of information available and help guide prevention efforts.

# The Use of Data and Research in Comprehensive Fire Prevention Programs

A comprehensive prevention effort includes compiling and analyzing data. That data can be statistical or anecdotal in nature, and both types of data produce valuable information that aids in the design of proactive prevention strategies. Whether conducted by investigators or analysts, this function of an effective fire investigation effort is critical in helping to design preventive solutions. Analysis can be done anecdotally on specific fires or on a large scale with databases containing information on hundreds of fires. The National Fire Incident Reporting System (NFIRS 5) provides a template for collecting data in a report format. Gathering the data is important, but analyzing it is just as critical and can sometimes require specialized expertise. This type of research is important to make sure that prevention programs are appropriately designed and targeted to the intended audiences and to the problems fire departments are trying to mitigate.

Another type of research comes from everyday occurrences. Anecdotal evidence about human behavior and other contributing factors can be especially revealing. Examination of a single fire can provide a good example that is understandable to decision makers unfamiliar with the technical aspects of fire cause. A thorough investigation might reveal that papers left too near a portable heater, doors left open to let air feed the fire, and the flammability of wall coverings *all* contributed to the fire cause and its rapid advance. An analysis of overall data can reveal the types of behaviors and the occupancies that produce the most frequent fires. Specific examples such as these can bring the complex nature of investigations and data analysis to a simple level that lay people can readily understand.

In addition to anecdotal evidence, long-range evaluation of loss data should be part of the goal of an effective prevention program so that proactive strategies benefit from the historical perspective a good investigation and analysis process provides.

Consequently, the record-keeping system for fire investigations and overall losses is a critical long-term tool of comprehensive prevention efforts.

Sometimes national research is available to help shape prevention efforts when local data is not available—or when resources for data collection and analysis do not exist. Under these circumstances, programs can be obtained or designed where there is some hope of linking the effort to meaningful data that helps target our efforts. And the lessons we learn from research on large fires, such as the Station nightclub fire in Rhode Island, are an important element in our local and collective efforts. Many jurisdictions across the nation shaped their fire sprinkler requirements to increase protection in nightclubs as a result of the investigation, research, and report of findings for that particular fire.

## EVALUATING COMPREHENSIVE PREVENTION PROGRAMS

The major parts of a comprehensive prevention program include engineering, enforcement and educational activities that help to reduce risk and losses within a community. However, there are some underlying management concerns that must be addressed. Specifically, local decision makers are asking how prevention efforts can be evaluated to ensure that they are being managed effectively and efficiently. These decision makers may be fire officials or those outside the fire service with an interest in the effective management of public resources.

Local decision makers are facing more pressure to justify expenses for every type of government service. Concerned taxpayers want to know what results are produced by their tax dollars and whether programs are managed efficiently. Prevention programs are often the most difficult to evaluate, even though most taxpayers accept the idea that preventing an incident is cheaper than dealing with it after the fact. However, there are performance measures available that indicate whether prevention programs are producing the desired results and doing so in an efficient manner.

The Governmental Standards Accounting Board has developed evaluation measures that can be applied to fire department activities, though the board stipulates that it is very difficult to compare one jurisdiction with another. This is because the variables in each jurisdiction are difficult to match. The U.S. Fire Administration has also produced measures in conjunction with California Polytechnic Institute. Generally speaking, the combination of these indicators can be categorized as workload, efficiency, or effectiveness measures.

*Workload measures* are those that document the amount of work conducted. These measures include topics such as the number of code enforcement inspections done per inspector or the number of presentations by public educators. *Efficiency measures* are those that demonstrate whether something is done quickly and at the lowest possible cost. These measures would include the cost per inspection or the cost per public education presentation. Effectiveness measures may produce the most solid results to local decision makers when evaluating prevention efforts. *Effectiveness measures* show the impacts or outcomes of specific prevention efforts and their relationship to the stated goals and objectives of those efforts. Effectiveness measures get at the heart of the question usually asked by concerned taxpayers: Why is this service in place, and what is it providing us?

The effectiveness measures most commonly used include educational gain, risk reduction, and loss reduction. Measuring educational gain provides evidence that public fire and life safety educational activities are producing a desired learning result. Pre- and posttesting practices can document whether the recipients of an educational program are actually learning or merely sitting through a presentation they will forget the next day.

Managing risk is one of the basic reasons for having a comprehensive prevention program. Measuring changes in that risk means documenting effects such as an increase in safety behaviors or a decrease in hazard-producing behaviors. For example, documenting the number of working smoke detectors in a community can provide evidence that the risk of dying in a fire is reduced. National statistics indicate the effectiveness of smoke detectors in saving lives. A compilation of hazards abated during fire inspections can also provide evidence that risks have been reduced because hazards have been removed.

Measuring risk reduction is a valuable part of an evaluation strategy because it provides some quantifiable indicators that can be used to determine the impact of a prevention program. For example, a random sample survey of citizens in a community might indicate how many are practicing safe behaviors that *lead* toward a reduction in fire deaths or property loss. Local decision makers can quantify how many people have working with smoke detectors and how many practice fire escape planning. They can also quantify how many community fire hazards have been abated in a code enforcement inspection program.

However, the ultimate performance measures will always be those that document loss reduction. Workload measures will demonstrate that employees are doing an adequate amount of work. Efficiency measures will provide some indication of how quickly things are done or what they cost in relation to other similar services. But measuring the reduced losses that occur as a result of prevention programs will provide the strongest evidence of positive results. Risk reduction measures are the ultimate performance (or outcome) measures that justify the expenses of conducting prevention programs. However, local decision makers should be cautioned against leaping to conclusions based on short-term analyses of loss data. *All* the performance measures looking at effectiveness should be evaluated over a period of time, because a change in activity may be caused by normal variations in any statistical analysis. Looking at loss reduction over a period of time can provide the best picture of whether a local jurisdiction is improving or not. In addition, it is more accurate (statistically) to compare a jurisdiction to its own history, rather than trying to compare it to another.

More recently, the National Fire Academy, the National Fire Protection Association, and others have been using a slightly different array of prevention performance measures to indicate their results. Most commonly, they include formative, process, impact, and outcome measures. The *formative measures* are those that indicate the level and type of research done to develop prevention programs. *Process measures* reflect the benchmarks of progress for a particular program, or capture data such as the workload so that the progress of efforts may be measured. *Impact measures* are usually those that successfully mitigate risk—such as documenting educational gain or elimination of hazards. Finally, *outcome measures* are those that document changes in the losses generated by fire and/or injuries. The economic (direct and indirect) losses and changes in the injury levels or deaths are all part of the outcome our prevention efforts are designed to produce.

More is said about evaluating prevention programs in Chapter 12.

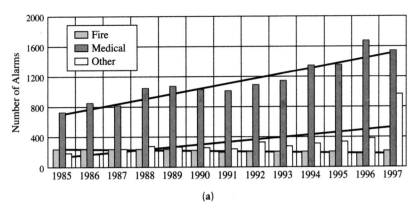

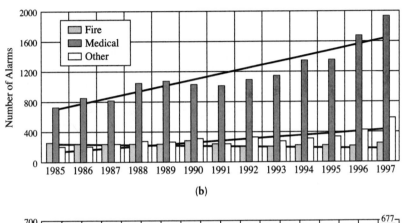

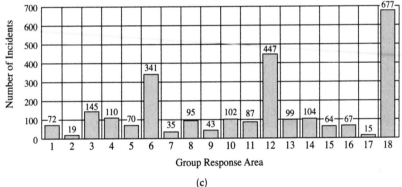

(a), (b), (c)   An example of how the results of fire prevention efforts can be tracked and presented. A trend analysis of fire incident rates over a period of time.

# Administrative Role of the Fire Marshal

Clearly, the person responsible for managing a comprehensive prevention effort should be highly trained and qualified. Understanding all the concepts listed in this chapter alone is daunting enough. Gaining expertise in all of them may be impossible in a normal career. That is why the professional qualification standard for fire marshals (NFPA 1037) was designed as a generalist and administrative job performance document. In smaller departments, where only one or two services are offered and personnel are limited, getting training as a technical practitioner (like a code enforcer or fire investigator) is probably more important. In those cases, other professional qualification standards would apply. But departments of any

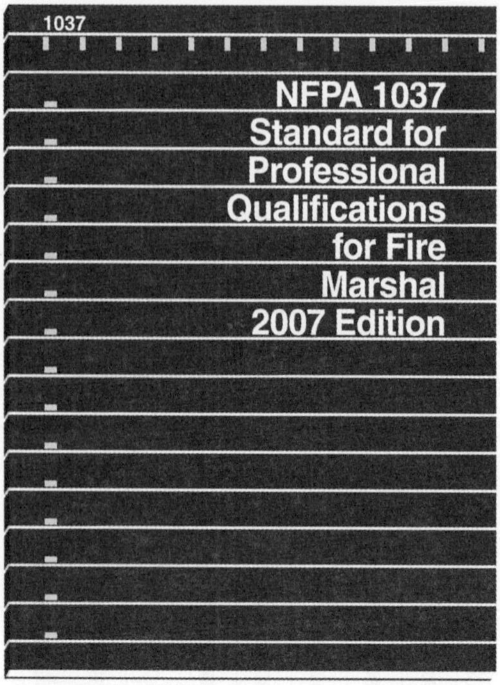

NFPA 1037 is a professional qualification standard that outlines the administrative duties of the fire marshal, as well as the job performance requirements for managing a comprehensive approach to fire prevention efforts. *(Reprinted with permission from National Fire Protection Association, © 2008 NFPA)*

**1037**

**NFPA 1037**
**Standard for**
**Professional**
**Qualifications**
**for Fire**
**Marshal**
**2007 Edition**

administrative duties of the fire marshal
■ the management aspects of the job, rather than the technical expertise of any given portion of it

size offering a variety of prevention services need a professional capable of carrying out the **administrative duties of the fire marshal** and understanding how comprehensive prevention efforts fit together.

Fire marshals should know that effective fire prevention programs are more than *community relations*—though such programs include good relationships as a foundation for success. Fire marshals need to understand that using all the tools of engineering, enforcement, education, and fire investigations and then measuring their impacts and outcomes is what managing fire prevention programs is all about. Putting all the pieces together for a comprehensive approach to fire prevention requires an effective administrator, no matter the size of the effort, because multifaceted approaches to prevention are proving to be the most effective at producing results.

## Summary

Comprehensive prevention programs are a combination of loss-control strategies designed to produce results in reducing risk or losses in any given community. They include elements of engineering for safety solutions, such as the plan review of new construction for water supply, fire department access, and fixed fire protection systems. They include enforcement efforts, such as well-designed fire inspections programs that ensure compliance with a properly adopted and administered code enforcement program. They include educational activities that are well designed to reach the public through the school system or by other targeted means. They include an adequate fire investigation program that helps guide the development of effective prevention strategies by providing the information necessary to target prevention efforts appropriately. Comprehensive prevention programs also include efforts to help control the problem of arson in their community.

Comprehensive prevention programs can be carried out by specialized personnel, emergency response personnel, or even volunteers. They can involve any element that includes the public in an active role to better protect themselves, as in self-inspection programs for code enforcement or through neighborhood coalitions that go door to door with appropriate fire and life safety materials.

However they are conducted, comprehensive prevention programs are multifaceted, using the full variety of prevention strategies that all work toward the common goal of prevention—fewer deaths and less loss. The future will make even greater demands on the level of expertise and training these people must attain as the technological and legal issues facing them become more complex. In addition, greater emphasis on performance-based codes, the movement toward all-injury control, and the establishment of coalitions to promote fire and life safety are just some of the emerging trends that prevention programs will face in coming years.

## Case Study

### Wareville Fire Prevention Efforts

Wareville is a medium-sized coastal community with a population of just under 150,000. Their prevention programs are briefly described by category in the following text.

#### EDUCATION PROGRAMS

Public education programs are developed and delivered by specialists or by other inspectors assigned to the Prevention Division. Educational efforts include the following activities:

- *Follow the Footsteps to Fire Safety* is a home-grown fire prevention program for young children and their parents that is designed to lead them toward safe behaviors for a variety of fire problems.

- Child fire-play intervention is provided to children who have set a fire and their parents. Screening, educational intervention, and referral services are all part of the program.
- *Risk Watch*™ is the general injury prevention program being used by Wareville. It is taught by specially trained fourth-grade schoolteachers, who are in turn trained by the department's public education staff.
- *Stay Cool!* is a locally developed program that teaches firesetting prevention to fifth graders.
- Fire evacuations are planned and practiced in all schools during Fire Prevention Week and in any other occupancy on request. All high-rise buildings and apartment buildings are required to have an evacuation plan that is provided to the building's tenants.

- *Fire Wisdom* is another locally developed program that teaches fire safety behaviors to seniors.
- Fire prevention staff also regularly participate in community health fairs, distributing fire and life safety information.
- Fire prevention staff provide neighborhood safety camps each summer. Children participate in reporting a fire on a simulated 911 system, identify home fire hazards, practice stop-drop-and-roll techniques, and evaluate home fire escape routes. They also try on firefighter's turnout gear and use fire hoses as a reward for their efforts.

## ENFORCEMENT (CODE COMPLIANCE) PROGRAMS

Wareville has a staff of five inspectors and one inspector supervisor. Two are certified hazardous material technicians. Most are certified fire inspectors.

- The Certificate of Occupancy renewal program is the major proactive code compliance mechanism for Wareville. Each building that needs a certificate of occupancy requires periodic inspections and compliance for renewal of the certificate. Renewals are required annually for high-hazard occupancies, every 2 years for warehouses and assembly and institutional occupancies, and every 3 years for mercantile, educational, and business occupancies. Apartment buildings are on a 1-, 3-, or 5-year cycle, depending on the condition of the building. Fees are charged for inspections, which recover the total cost of the program.

However, cutbacks in staffing in recent years have lengthened the inspection cycle for all properties.

- Between the Certificate of Occupancy renewal periods, complaint-based inspections are performed.
- Permits are inspected and issued for fireworks, explosives, and open burning.
- Day care and foster care homes are inspected for fire safety as a condition of their licensing.
- Hardwired smoke detectors are required as a condition of the sale of a home. This is enforced by the Fire Prevention division.

## PLAN REVIEW

The plan review process is staffed by a fire protection engineer and a fire sprinkler inspector.

- Plan reviews are performed of fire sprinkler systems, fire alarm systems, smoke removal systems, and any other fire-related designs.
- Inspections are done with the contractors for final approval of sprinkler, fire alarm, and smoke removal sytems.
- Water supply flow testing is performed by the fire protection engineer to determine the amount of water available for firefighting and built-in suppression systems.
- The backlog of plans to review for new construction is a continuing source of aggravation for the mayor and city manager—and complaints about the process are more frequently being sent to the fire chief for action.

# Case Study Review Questions

1. What elements of a comprehensive prevention program are covered by Wareville's programs?
2. Given the relationship between fire investigation and prevention efforts, what recommendations would you make to the Chief of Wareville about their efforts?
3. Looking at comprehensive prevention programs in general, how would you say the movement toward all-injury control will affect efforts for plan review, code enforcement, and investigation? In a coastal community, should something be done about the dangers of tsunami?
4. Why should a fire department be interested in the plan review process, and what fire and life safety concerns may be addressed there?
5. How can data be used effectively to design prevention programs?
6. What kinds of things does a fire department usually look for in the plan review process?

7. How does fire department access become critical for wildland fire risk areas?

8. What is the difference between administering the fire code and enforcing it?

9. What does an appeals process do for individual inspectors?

10. What are the advantages of an all-injury safety curriculum for schools?

PEARSON
## myfirekit™

For additional review and practice tests, visit **www.bradybooks.com** and click on MyBradyKit to access book-specific resources for this text! Your instructor may also assign Additional Project work related to topics in this chapter.

Register your access code from the front of our book by going to **www.bradybooks.com** and selecting the mykit links. If the code has already been scratched off, go to **www.bradybooks.com** and follow the MyBradyKit link from there.

# Endnotes

1. See TriData report, *Proving Public Education Works*. Arlington, VA: TriData, 1990.
2. *Risk Watch*. Quincy, MA: National Fire Protection Association.
3. Hall, John R., Jr. *U.S. Fire Death Patterns by State*. Quincy, MA: National Fire Protection Association, Fire Analysis and Research Division, 2000.
4. See TriData report, *A View of Management in Fire Investigation Units,* vol. 1 and 2. Federal Emergency Management Agency and the U.S. Fire Administration, date.

# 2

# Risk- and Loss-Reduction Educational Programs

## KEY TERMS

Coalitions and partnerships *p. 45*
Combined prevention programs *p. 44*
Evaluation *p. 35*

Public information and public relations *p. 28*
School-based educational programs *p. 38*

Steps in public education planning *p. 30*
Targeted educational programs *p. 41*

## OBJECTIVES

After reading this chapter you should be able to:
- Differentiate among public education, public information, and public relations programs.
- List elements of a planning process specific to public fire and life safety education programs.
- Describe the difference between school programs and other targeted educational programs.
- Explain why coalition development and community involvement in prevention efforts are important.
- Explain the links among arson prevention, public education, and other community fire prevention efforts.

This chapter will deal with the topic of fire and life safety education programs in some depth, and there are a number of references in the text to quality materials that can help produce effective programs. Some elements are critical to *all* educational efforts if they are to be effective. These are the basic building blocks that fire officials and other local decision makers should understand to help direct their public education programs toward an effective outcome.

# The Differences Among Public Education, Public Information, and Public Relations Programs

Public fire and life safety education programs are a critical part of any comprehensive prevention effort. In marketing terms, they fill a specific niche in the mix of prevention "products" used to control losses for any community. They are important because most fires and fire deaths occur in residential occupancies, where the ability to require engineered safety features or to enforce fire safety behaviors is often restricted and nearly always resisted. The issue of affordable housing is so important to most communities that adding costs for fire and life safety features is usually met with heavy resistance from the building and development community.

Home builders usually resist new safety features in homes, particularly residential fire sprinklers, as evidenced by recent battles in the code promulgation arena over mandating their use in one- and two-family dwellings. Political leaders are sensitive to the issue of cost and are not easily persuaded that proven safety features such as fire sprinkler systems are a necessity. They are often motivated to provide housing that is affordable to the largest portion of the population they serve. And during tough economic times, the pressure to keep construction costs low is even more intense.

In addition, the home (even when it is an individual apartment unit) is viewed as private property and not subject to government intervention, except in rare circumstances and then with due legal process procedures followed. Because of these facts, public fire and life safety education efforts are necessary if the fire service is to have any significant impact on their fire loss statistics.

This chapter deals in some depth with the topic of fire and life safety education programs, and there are a number of references in the text to quality materials that can help produce effective programs. Some elements are critical to *all* educational efforts if they are to be effective. These are the basic building blocks that fire officials and other local decision makers should understand to help direct their public education programs toward an effective outcome.

Among the items that fire officials and local decision makers must understand about public fire and life safety education programs is the difference between public education and **public information and public relations**. They should also understand the basic planning process for effective programs, the base programs in schools, the many targeted programs that may exist, the interrelationship between educational efforts and other prevention strategies, the movement to establish coalitions and involve the community in public education programs, and how to evaluate public education efforts.

Mary Nachbar-Corso, formerly the State Fire Marshal of Washington, described in an edition of *The Fire Chief's Handbook*[1] what she called the "accident attitude theory." It describes a cultural value about accidents that permeates our modern industrial society, creating a false impression about emergency events that can be mitigated. She says:

> In reality, it might be said that there is no such thing as an accident when it comes to most fires, injuries, or deaths. *There are only caused occurrences.* If you disagree with this conclusion, consider this premise: in Webster's dictionary, we find the word, "accident" to mean "an event occurring by chance." Winning the lottery is an accident. Having a fire is not. When the words "accident" or "accidental" are included in fire incident reports or told to the media, we perpetuate the myth that the fire was an unavoidable event. Nearly every fire is the result of carelessness, heedlessness, thoughtlessness, or ignorance on the part of someone. It is not a cigarette that falls asleep while someone is smoking it. The events we commonly refer to as accidental are really controllable and avoidable. This is a harsh reality, yet true.

Nachbar-Corso has done an excellent job of framing the issue that public fire and life safety education efforts must address. Education is designed to raise the cognitive level of an individual or group so that they understand the message. But it must ultimately change their attitude about the topic so that they will actually change their behavior. The goal of educational programs is to change behaviors, which reduces the number of incidents or, at a minimum, reduces the risk of those incidents.

Awareness or public relations programs are designed to make the audience *aware* of the message. An example would be an informational campaign designed to enhance the image of the fire service. Audience members may be able to repeat the message, and even adopt a feeling as a result. But a public information or relations campaign is not purposely designed to bring about a specific behavioral change.

Consider a marketing parallel. Marketing campaigns must first make the potential customer aware that the product exists. A public information campaign would make a customer aware that a product exists—but professionals in the field tell us that more in-depth efforts, repetition of messages, and market saturation are what it takes to get people to act on the message they receive. In the final analysis, marketing efforts are only successful if those customers actually buy the product. The end result is an action on the part of the intended audience. For public fire and life safety education efforts, the end result must be measured in the same terms. And in both cases (marketing and education), producing behavior change requires more money.

If we want to produce results (an impact on our outcome) with our educational efforts, it is not enough to merely present information. Successful educational programs raise the cognitive level of understanding and *help motivate the audience to take action* based on that information.

## EDUCATIONAL ELEMENTS

**AFFECTIVE LEARNING DOMAIN:**

The portion of the learning process related to attitudes. Someone unwilling to learn won't no matter how good the educational offering. Changing attitudes and opening people up to learning is therefore a critical part of the education process.

**COGNITIVE LEARNING DOMAIN:**

The portion of the learning where cognitive abilities change and can be documented. Being able to recite newly learned material—or to write it—are signs of cognitive learning increases.

**PSYCHOMOTOR LEARNING DOMAIN:**

The portion of the learning process where physical skills are learned. Being able to cognitively recite learned material is one level of learning. Being able to repeatedly perform it requires accessing the psychomotor learning domain—and achieving the skill set necessary to physically perform.

The National Fire Academy has developed a number of courses that provide training for potential public fire and life safety educators. In each, they have addressed the basic foundations of learning that delineate the differences between public information and educational strategies. In simple terms, those elements that make up the foundation of education include the cognitive, affective, and psychomotor learning domains.

When related to educational efforts, these learning domains can be viewed as a progression of the depth of understanding an audience must achieve if it is to act on the information. The cognitive domain refers to the audience members' ability to "know" and understand the principles being taught. The affective domain refers to their attitude toward that information. The psychomotor domain refers to their ability to perform the task being taught or to act on the information. Put more directly, these learning domains mean that audiences must be taught the proper message, and they must be sufficiently motivated by the message to change their attitude toward it so that they will take action. They must also actually be able to perform the action.

An example can be drawn from the simple message "stop, drop, and roll." Many are taught that if their clothes catch fire, they should "stop, drop, and roll" to smother the flames. An informational campaign would make them aware of the message. An educational program would make them aware of what to do, convince them that they should actually do it, and ensure that they have an understanding of the physical requirements necessary to perform it. Most education experts agree that getting someone to see, hear, and perform the task is the best way to ensure that they will learn it well enough to be able to do it.

Public relations campaigns, much like informational programs, are not designed to change behavior. They are designed to ensure a positive image and to foster relationships. For example, we would want taxpayers to understand the nature and value of our services, which is a public relations or information goal. But reaching the deeper cognitive level combined with behavior changes is not considered part of public relations.

In summary, the difference between informational and public relations programs and educational programs is the depth of learning. The final result of educational efforts must be to motivate audiences strongly enough to get them to take action.

In our society, motivation presents a significant challenge. As Nachbar-Corso states, our culture still (largely) views the events that produce fires or other injuries as accidents. Until people understand that these events usually occur because of a predictable pattern of behaviors, and until they are sufficiently "educated" so that their attitude changes, real action will not take place. It is possible to change behavior, and there are many examples of successful educational programs that use modern techniques to produce results.

TriData Corporation, a division of System Planning Corporation of Virginia, has conducted numerous studies on the topic of public education and produced a guide that provides examples called *Proving Public Fire Education Works*.[2] Rossomando and Associates has also produced guides that help explain the necessary components of an effective education strategy. *The Community-Based Fire Safety Education Handbook*[3] and the guide called *Reaching High-Risk Groups: The Community-Based Fire Safety Program*[4] use several anecdotal examples to show that effective educational strategies can be developed and produce measurable results. Among the critical factors for development of effective educational programs is the planning process used to design them.

# Planning for Educational Programs

In the early 1980s, the U.S. Fire Administration developed a planning process specifically for public education programs. It is as relevant today as it was then. Called the Five-Step Planning Process, it outlines steps that may be followed specifically or more generally in conjunction with the strategic planning process outlined in Chapter 7. The **steps in public education planning** include *identification, selection, design, implementation,* and *evaluation.*

The International Fire Service Training Association (IFSTA) produces an excellent manual for public education programs that outlines this planning process in greater detail.[5] It should be noted that this planning process is continuous and not a one-time event. Those using the process should also be mindful that the entire plan is completed *before* any attempt to implement it is begun. In this way, planners have an idea about how the entire plan fits together before moving forward with any one segment.

What follows is a simple description of the planning process that is also incorporated in Chapter 7 for strategic and tactical planning for all prevention programs. The steps listed here are more specifically oriented toward public education programs.

## IDENTIFICATION

The step that deals with identification refers to the process of gathering enough data to identify which problems will be targeted for specific educational programs. It is also called formative evaluation and is dealt with again in Chapter 12.

**steps in public education planning**
▪ *Identification* refers to identifying problems we need to address through research; *selection* refers to selecting achievable and specific goals; *design* refers to designing programs toward those goals and specified audiences; *implementation* refers to development of an implementation plan; and *evaluation* refers to development of an evaluation plan. All five steps are completed before implementation actually begins.

The identification stage is particularly important when there are not enough resources to conduct programs for each type of problem a fire department faces. It is also critical when developing targeted programs designed to reach a particular audience with a specific message.

Gathering fire loss or other emergency incident information is the first part of the identification phase. Some fire departments must use general information or make some informed assumptions because their loss data is not detailed enough to provide specific direction. The goal of this part of the process is to identify problem areas specifically enough to create a profile that outlines high-risk locations, times, victims, and behaviors. For example, an analysis of available loss data might lead one to conclude that the principal fire problem for a community is careless cooking practices. It might further pinpoint the problem as occurring in a particular part of town, during the after-school hours, and the principal audience involved in the behavior as teenage girls. The so-called latchkey children, who are responsible for themselves until parents arrive home from work, could potentially be a big part of the problem for cooking fires. The actual research done to determine who is at risk and what they are doing to cause that risk takes a good deal of effort, but it will provide evidence to allow targeting of problems and audiences rather than making uninformed assumptions. This formative (planning level) research is a critical part of making sure public education messages are on target.

In actual practice, an analysis of available loss data usually produces a much more cluttered picture of the problems facing a local fire department. For example, many fire departments are now questioning their prevention strategies based on an analysis of the data, because they are discovering that most of their emergency incidents are medical in nature. The process of identifying real problems has created the movement toward *all-injury control* that many fire departments are embracing as a reality for their prevention strategies. This means that they are beginning to apply the same basic tools toward programs designed to prevent or mitigate losses from injuries caused by incidents other than fire.

The identification phase also provides an opportunity for the fire service to evaluate the potential problems they may face as well. The real (statistical) problems they deal with may be less significant when compared with the potential loss a particular type of building represents. For example, hospitals traditionally have a lower fire incident rate, but the potential exists for mass casualties if one should occur. Consequently, most fire services conclude that even if the statistical probability of hospital fires or other emergencies is low, they must be planned for because of the potential for disaster they represent.

The identification phase of the planning process is critical to help identify specific targets and direct efforts toward their most effective outcome. The next phase, selection, is equally important to help filter the decision-making process toward a logical target.

## SELECTION

Analyzing loss data will produce several possible targets for educational programs based on real or potential problems that are revealed through that review. The selection step of the planning process means choosing from among many problems to decide which will receive attention. At this phase of the planning

These trends in various emergency incident call volumes represent a principal source of data that forms the foundation for evidence-based decision making.
(*Courtesy of Vancouver Fire Department, WA*)

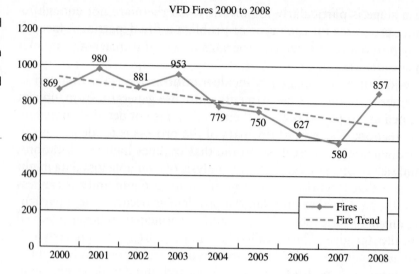

VFD Fires 2000 to 2008

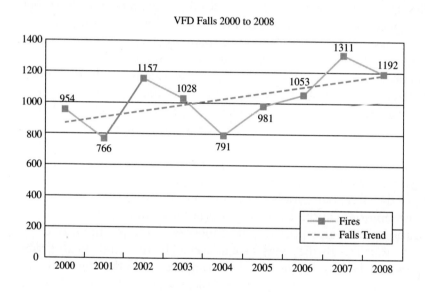

VFD Falls 2000 to 2008

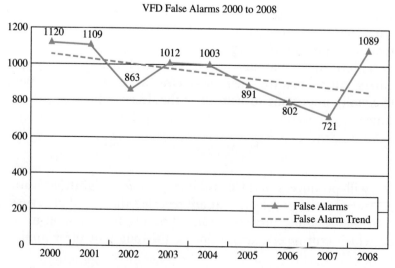

VFD False Alarms 2000 to 2008

process, fire officials and other local decision makers have the ability to direct their efforts toward the populations and problems they feel will provide the highest return on their investment. Put another way, they can "get the most bang for their buck" by targeting their efforts. More is said later in the chapter about the relationship between targeted and broad-based educational programs.

However, it is important for those responsible for developing effective educational strategies to do so in a thoughtful fashion. It is during this phase of the plan that planners begin to direct their efforts so that the program can be designed appropriately for the identified problem and audience. At this point of the planning process, some kind of cost-benefit analysis should also be done to help determine the most appropriate course of action. Specific objectives should also be identified so that the program results may be measured. It is during this phase that the problems are identified in specific terms so that an educational program can be built around the premise that one knows or can logically infer who is at risk, what behaviors are causing that risk, and where those at risk can best be reached. In simple terms, this phase narrows the "who, what, when, where, and how" of problems to a manageable size.

## DESIGN

The design phase of the planning process takes the information from the first two phases and begins to organize it in a fashion that will most likely achieve results. The principal factors that determine the design of the educational strategy are related to who is at risk, what they are doing to cause that risk, and where they can be reached. It is here that educators must determine the proper message content, format, time, and location and give some thought as to how it will be tested before implementation begins.

Message content really means what we are trying to say. "Stop, drop, and roll" is an example of message content. Professional marketers tell us that these messages should be short, positive, and to the point. In the education arena, they should tell people what to do, rather than what not to do.

Message format refers to the type of delivery system used to convey the message. Radio, television, newspapers, classroom presentations, Web pages, billboards, and presentations to business groups are all examples of different formats used to convey the message. Those responsible for developing educational programs should relate the message content and format to the high-risk audience, location, and time. For example, it is not effective to try to reach a target audience of senior citizens with a puppet presentation in the schools. It is more appropriate to reach them in senior centers or through targeted media outlets that cater to the elderly with a type of message and presentation that will appeal to individuals in their age group.

Consequently, the design phase of the planning process is where the message and format are matched to the audience and their location. It is critical in this portion of the plan that considerable thought be given to the type of message that will motivate the audience to act on the message. Reaching adults by appealing to their sense of protection for their children may produce positive results. Using a senior spokesperson to talk to other seniors may also be effective. The key is identifying what is important to the intended audience and using that information to help convey the importance of the message.

During the design phase, it is important to consider whether there is any product already in existence that matches the intended message and audience.

Home Safety Literacy Project

Many involved in education make the mistake of thinking they must spend time in development, when some product already exists that meets their needs. Educators must determine if it is better to spend time in development or delivery of the product. If development of a new product is necessary, it is extremely important to field test it before it is implemented on a large scale. Many public fire educators have found that their own assumptions about the quality of a message are incorrect after it has been reviewed by the intended audience.

In addition, it is not unusual to hire professionals in the design field to handle this portion of the planning process. It would take years to duplicate the type of training that many marketing professionals receive. Sometimes, it is less time-consuming and more effective to hire professionals to help design and even implement public education strategies.

## IMPLEMENTATION

The implementation phase of the planning process deals with the procurement of materials and resources to carry out the plan, as well as actually scheduling its delivery. If physical presentations are to be used, then scheduling the presenters and providing them with the proper training and material for delivery will all be factors that must be dealt with in this phase of the plan. Using radio or television advertisements will mean developing quality ads and purchasing air time for their display to the intended audience. In other words, the implementation phase is where the logistics of the program are identified and planned in detail.

A Gantt chart is a tool often used to help organize projects and provide an implementation schedule and a list of tasks to be done. It is also helpful in tracking who is responsible for those tasks. Depending on the size of the project, some

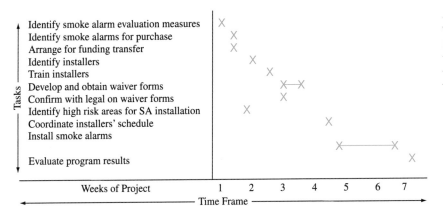

A Gantt chart—in whatever form—can help track issues and tasks in an organized fashion.

kind of charting is valuable in keeping an implementation plan on track. But it is important to remember that development of the implementation plan occurs *before* a plan is actually implemented.

## EVALUATION

The **evaluation** phase of the planning process is where decisions are made about how the program will be monitored and measured. It should be obvious that those responsible for the management of public resources want to be able to see the results they produce. Consequently, fire officials should be looking at benchmarks and measurable outcomes from educational programs that will demonstrate their effectiveness.

**evaluation**
■ the gathering and analysis of data (statistical or anecdotal) that provides evidence of a program's impact or outcome

More is said about evaluating prevention programs in Chapter 12 of this book. However, there are portions of the overall evaluation strategy that lend themselves more specifically to educational programs. The goal of educational programs is to raise the knowledge level of the audience and motivate them to act on that information. Likewise, the evaluation measures should demonstrate that knowledge levels have indeed been raised and that the increase is a result of the educational program. It should also document which target audience behaviors have been changed as a result of the program.

Educational gain can be measured and attributed to a specific program by pre- and posttesting intended audiences. The gains from the pretest to the posttest can demonstrate that people are actually learning something, rather than merely sitting through a presentation. Educational gain can be measured individually, in the classroom, or even throughout an entire community. For example, the impact of an educational campaign for kitchen safety can be measured by pre- and posttesting in a classroom or by a pre- and postsurvey for a selected portion of the residents of the larger community. More is said about how pre- and posttests, or surveys, are developed in Chapter 12.

Ultimately, a test or survey must demonstrate that behaviors have changed as a result of the educational activity. Sometimes this can be observed in the classroom setting, but other times a phone or physical survey of the community is necessary to provide hard evidence of that behavior change. Voluntary home safety inspections provide an example of how a physical survey of hazards may be observed and documented. Documenting the fact that the intended audience acted on the information they received will demonstrate the risk-reducing behaviors brought about by the educational program. So the sequence of education, from increased knowledge (cognitive) toward changing attitudes (affective) and actually causing action (psychomotor), will be demonstrated by measuring the educational gain and showing the behavior changes that have occurred as a result of its effort.

A pre- and posttest, or survey, such as this is an important part of documenting the impact of a public education program.

---

**Smoke Alarm Survey**

Disconnected ☐
No Answer (Date) ☐ ☐ ☐ ☐
Decline ☐

Interview Number _____
Respondent Name _____ Phone Number _____
Interviewer Number _____ Date/Time of Interview _____

(ASK TO SPEAK TO HEAD OF HOUSEHOLD)

"Hello, this is _____ from the Oregon Fire Education Association. We are doing a study to determine public usage of residential smoke detectors. Would you help us by answering a few questions? Your answers will be strictly confidential."

1. "Do you have any smoke detectors in your home?"
   Yes . . . . . . . . . . 1
   No . . . . . . . . . . 2          IF NO, SKIP TO QUESTION 4          (5)

   A. "How many detectors do you have?"
      1   2   3   4 or more          (6)

   B. "Where are they located?"  DO NOT READ LIST:
      Hallway . . . . . . . 1          Basement . . . . . . 6          (7–8)
      Bedroom . . . . . . 2          Kitchen . . . . . . . 6
      Living Room . . . 3          Other (specify) . 7
      Stairwell . . . . . . 4

2. "To the best of your knowledge, is your detector in operating condition?"
   Yes . . . . . . . . . . 1          No–Battery removed . . . . 4          (9)
   No–Unplugged . . 2          No–Det. taken down . . . . 5
   No–Bat. Dead . . . 3

3. A. "How often do you or someone in your household test your detectors?"
   DO NOT READ LIST
      Do not test . . . . . . . . 1          Every 6 months . . . . . 6          (10)
      At least once a week . 1          Once a year . . . . . . . 7
      Twice a month . . . . . . 2          Other _____ 8
      Once a month . . . . . . 3                    (Specify)
      Every 3 months . . . . . 5

   B. "How do you test your detector?"  DO NOT READ LIST
      Real smoke . . . . . . . . 1          Check light . . . . . . . . 3          (11)
      Push button . . . . . . . 2          Other _____ 8
                                                      (Specify)

4. "What would you say has prevented you from purchasing a smoke detector?"
   DO NOT READ LIST.  IF SEVERAL REASONS ARE GIVEN, CIRCLE ALL THAT APPLY.  PROBE FOR MULTIPLE REASONS BY ASKING: "Any other reason?"
      Not available locally/can't find them . . . . . . . . . . . . . . . . . . . . 01          (12–13)
      Too expensive . . . . . . . . . . . . . . . . . . . . . . . . . . . . . . . . . . 02
      Don't consider them necessary or worthwhile . . . . . . . . . . . . 03
      No interest/never thought about it . . . . . . . . . . . . . . . . . . . . 04
      Keep forgetting . . . . . . . . . . . . . . . . . . . . . . . . . . . . . . . . . 05
      Can't install . . . . . . . . . . . . . . . . . . . . . . . . . . . . . . . . . . . 06
      Detectors aren't reliable . . . . . . . . . . . . . . . . . . . . . . . . . . . 07
      Other (specify) _____ . . . . . . . . . . 08

*(continued)*

"Now, I would like to ask you a few questions for classification purposes only:"

5. "Do you own or rent your residence?"
Own . . . . . . . . . 1          Rent . . . . . . . . . . 2                    (14)

6. "What type of residence do you live in?"  READ LIST IF NECESSARY
One- or two-family dwelling . . . . . . . . . . . . . . . . . . . 1           (15)
Apartment, condominium, or townhouse . . . . . . . . . 2
Mobile home . . . . . . . . . . . . . . . . . . . . . . . . . . . . . 3
Other (Specify) _____ . . . . . . . . . . . . . 4

7. A. "Does your household have any children in elementary school?"
Yes . . . . . . . . . . . 1          No . . . . . . . . . . . . . . 2               (16)

B. "How about junior high or high school?"
Yes . . . . . . . . . . . 1          No . . . . . . . . . . . . . . 2               (17)

8. "How old is the head of household?  Please stop me when I come to the
appropriate age category."
18 to 24 . . . . . . . . . . . . 1          55 to 64 . . . . . . . . . . 5          (18)
25 to 34 . . . . . . . . . . . . 2          65 and over . . . . . . . 6
35 to 44 . . . . . . . . . . . . 3          Refused to answer . . . 7
45 to 54 . . . . . . . . . . . . 4

9. "May I ask what is your total yearly household income before taxes?  Please
stop me when I come to the appropriate income range."
READ LIST
Under $10,000 . . . . . . . . . . . . . . . . . . . . . . . . . . 1               (19)
$11,000 to $15,000 . . . . . . . . . . . . . . . . . . . . . . . 2
$16,000 to $20,000 . . . . . . . . . . . . . . . . . . . . . . . 3
$21,000 to $25,000 . . . . . . . . . . . . . . . . . . . . . . . 4
$26,000 to $50,000 . . . . . . . . . . . . . . . . . . . . . . . 5
$51,000 and over . . . . . . . . . . . . . . . . . . . . . . . 6
Refused to answer . . . . . . . . . . . . . . . . . . . . . . . 7

10. A. "Have you heard any announcements or information on smoke detector
maintenance and installation recently?"
Yes . . . . . . . . . . . 1          No . . . . . . . . . . . . 2                   (20)

IF NO, SKIP TO END
                                                                                 (21)
B. "Where did you get this information?"  DO NOT READ LIST
Radio . . . . . . . . . . 1          Landlord . . . . . . . . . 6
TV . . . . . . . . . . . . 2          Mail . . . . . . . . . . . . 7
Paper . . . . . . . . . . 3          Children . . . . . . . . . 8
Fire Dept. . . . . . . . 4          Other             9
Neighbors. . . . . . . 5               (Specify)

"Thank you for your cooperation."          _____

Because the purpose of educational strategies is to reduce risk and loss in the community, the effect of public fire and life safety education must also be determined by its impact on the fire and injury loss statistics for the intended audience. The educational campaign aimed at reducing the number of kitchen fires must ultimately be evaluated on its ability to reduce those types of fires.

It is important for those who develop public education strategies to know that all the steps of the planning process have been completed before the program is implemented. Even the steps for implementation and evaluation are presented as part of the planning process for a good reason: They must be planned for before they are actually done.

Effective public fire and life safety education programs are planned to achieve a measurable result. They must increase knowledge, change behaviors, and produce a measurable result if they are to be seen as effective. The most effective programs are planned; they do not happen by chance.

Not all programs are planned at the local level. Some are produced nationally to reach a broad audience and provide the basis for a minimum amount of knowledge that everyone must have if they are to be responsible citizens and reasonably safe. These programs must be planned as well, but are targeted to a broader audience. The best and most important examples of this type of program are those designed for schools.

# School Programs

Schools represent such a large portion of public fire and life safety education efforts that they are mentioned here as a basic staple of an effective prevention strategy. Unlike targeted educational programs, **school-based educational programs** are designed to reach a variety of audiences with the same basic messages. Thus, the objectives are broader, and the appeal to audiences must also be broader and appropriate for the age groups in schools.

Safety education programs in schools generally take two forms: physical presentations in the school or administration of a specialized curriculum designed for teacher delivery. Physically being present and getting students involved in activities has been demonstrated to provide an effective learning environment. Providing a specialized curriculum and letting the teachers do what they are trained for can also work well. Whichever approach is used, those involved with the delivery of public education programs must understand some of the basic concepts of educational methodology to design appropriate messages. It is important to keep that in mind whether we deliver educational messages in person or design programs to be delivered through teachers.

The most basic part of instruction is the communication flow between the instructor and the potential student. The communication feedback loop displays the pattern of communication between student and instructor, highlighting one critical point: It is a two-way communication process. Instructors cannot be sure that their message is being learned without getting feedback of one type or another. Pre- and posttesting is the surest way to document that learning has actually taken place; therefore, any effective public education strategy in the schools should incorporate at least some testing to validate its effect on students.

Another critical part of educational methodology is understanding how to design the product with specific educational objectives in mind. The National Fire Academy teaches that these educational objectives have four parts that must be considered when designing the program. They include the audience, the behavior, the conditions, and the degree of performance. Like the five-step planning process, these parts of the learning process play a part in the design of the final product.

The first thing to consider when developing the educational objective is the audience. All materials in the school program must be age, ethnicity, and gender appropriate if they are to be accepted by the audience. Those involved in the development of other school programs often specialize in their fields and should be consulted about the appropriateness of the material.

school-based educational programs
■ those programs that are general—covering several safety topics—and age appropriate. They may involve using a specialized curriculum where the teachers instruct their students, or presentations done in schools by fire safety professionals.

> **Educational Methodology**
>
> The elements of education
>
> AUDIENCE: The intended audience—sufficiently analyzed to reveal age and culture appropriate material decisions.
>
> BEHAVIOR: The specific behavior you want the audience to exhibit (e.g. install, inspect and test smoke alarms).
>
> CONDITION: The measurable condition you wish to achieve (e.g. installed smoke alarms that are inspected and tested) physically observed vs. phone survey.
>
> DEGREE: The degree to which you wish the audience to perform (e.g. 98% compliance rate for target audience.

The next part in creating the educational objective is to consider the behavior to be performed. As in the other elements of design for public education programs, the school-based program must consider the desired behavior and teach toward that goal.

The next portion of the educational objectives to consider is the conditions under which they will be performed. For example, if it is a specific task, such as "stop, drop, and roll," then the conditions would most likely require actual demonstration of the task. The condition is therefore the part of development that considers how the message will be taught. It is important to keep in mind that students usually best retain what they have seen, heard, and practiced.

The final portion of the objective is the degree of performance, or how well students must perform the task. For fire and life safety activities, the degree of performance is usually 100 percent for a given task such as "stop, drop, and roll." That is, each student must be able to perform the task exactly as demonstrated. In contrast, a written test of more general fire and life safety principles might have a passing score of 80 percent. The point is that consideration must be given for the expected student performance before designing the product, whatever form it may take. This is true whether it is a school presentation or a more broadly developed curriculum.

There is a great deal more to educational methodology than the relatively simple elements mentioned here. However, it is important for fire officials and other local decision makers to understand that school programs are not just put together by someone who speaks well in public. There is a science to education that teachers understand that must be considered before deciding which school programs are appropriate for specified students. This is why specialized curricula have become such an important part of the fire and life safety educator's toolbox.

## SPECIALIZED CURRICULA FOR SCHOOLS

There are a number of specialized safety curricula that have been developed for school children. The two most widely recognized are the *Learn Not to Burn*[®6] *(LNTB)* and the newer *Risk Watch*[®7] curricula developed by the National Fire Protection Association (NFPA). The *LNTB* program, as it is commonly known, uses teachers as the primary delivery mechanism to teach students fire safety behaviors. The success of the program has been documented for many years, both in terms of its educational soundness and also in terms of the number of lives it has helped to save. The NFPA responded to a demand from fire service and school professionals who are increasingly concerned about the fact that other emergencies produce more injuries and deaths than fire. They are also concerned about the increasing pressure on schools to meet community demands with diminishing resources. It has also become evident that fire departments are responding to more medical emergencies than fires.

As a result of these trends, NFPA developed an all-injury curriculum (*Risk Watch*), which uses the latest concepts in educational methodology. A critical part of the curriculum is that it presumes that teachers are still the best instructors. It uses activities that give students choices about proper safety behaviors and involves them actively in the educational process. Finally, it encompasses eight different injury types, including fire and scald prevention; firearm injury prevention; bicycle and pedestrian safety; motor vehicle safety; choking, suffocation, and strangulation prevention; poisoning prevention; falls prevention; and water safety.

*Risk Watch* promotes the concept of working with others to achieve common safety goals in the school system. By doing so, it draws attention to the fire safety issues of the fire service and incorporates a partnership of organizations that can make the most out of the school system with the least amount of disturbance to an already crowded schedule.

Among those partners who also have a concern about public safety are the police agencies, the medical community, and public health organizations. The Red Cross and the National Safe Kids Campaign are natural partners for this type of program, inside or outside the school system, and helped design the *Risk Watch* curriculum. Unfortunately, the *Risk Watch* curriculum is no longer promoted by the NFPA—but some limited support is available. They are focusing their efforts more specifically on the *Learn Not to Burn* curriculum.

There are other school curricula developed for safety purposes. For example, the Phoenix Fire Department has developed their own specialized version of an all-injury curriculum. The state of Oregon has also developed a version that incorporates tsunami and earthquake preparedness as part of the curriculum. However, those responsible for the development of a more localized product must put considerable effort into its design and testing. Those interested in developing their own product have an obligation to use scientific design methods, including modern educational methodology principles. In addition, using a coalition to support the effort is now a critical part of quality education efforts. Considerable time, effort, and expense together is involved in development. Consequently, fire officials must consider carefully when deciding whether to make a new product or to use one that already exists.

## SCHOOL PRESENTATIONS

Presentations within the schools are another basic staple of public fire and life safety programs. Many fire departments and private companies have developed clown shows, puppet shows, and other programs designed to appeal to a youthful audience. Some are entertaining, but few of these products have been tested to make sure that they adequately address the methodological issues of a proper

A public educator can provide educational presentations in schools alone—or combine the visits with a safety curriculum taught by the teachers. (*Courtesy of Portland Fire & Rescue, OR*)

educational program. They are usually most effective when used to supplement an established curriculum and provide the reinforcement necessary for a message to be retained over time. They also serve as an excellent reward for students and teachers who have covered the basics in the classroom setting.

In summary, school programs are a basic staple of public fire and life safety education efforts. They can take the form of individual presentations or more broadly based safety curricula. They are most effective when they combine elements of both, but any educational efforts in the schools must be designed well and take into account the modern educational methodology that ensures learning is actually taking place. Professional help in making the decision about design, purchase, and implementation of school programs is highly recommended.

# Targeted Educational Programs

However well they are designed, school programs will only cover part of the population that must get the fire and life safety message if efforts are to be effective. **Targeted educational programs** allow educators to meet the needs of specialized audiences.

Targeted programs are most effective when there is enough research done ahead of time to make sure that the educational approaches used will be appropriate and attractive for their intended audience. Targeted programs are often developed for specific problems and audiences, such as the program on security bars and fire safety developed by the Center for High Risk Outreach of the NFPA. The

**targeted educational programs**
- programs designed for specific messages and specific audiences

One proven effective way to distribute materials is door to door. This is an example of a simple topic and message designed for adults. (*Courtesy of Vancouver Fire Department, WA*)

Phoenix Fire Department has a well-designed program that deals with swimming safety and drowning prevention because their climate means that swimming pools are common, presenting a significant problem that might not be seen elsewhere. If a problem exists, it is safe to assume that a program has been developed to meet the need for education somewhere, but the same cautionary care must be given to product quality and appropriateness.

The U.S. Fire Administration (USFA) maintains a Web site (www.firesafety.gov) designed to serve as a resource directory of targeted programs produced locally or by private companies. The categories included on the site create a compendium of programs that list most of the specific problems encountered by any jurisdiction. The site includes programs for burns or scalds, cardiopulmonary resuscitation (CPR) and first aid, electrical hazards, fire escape planning, fire extinguisher use, flammable fabrics, residential fire inspections, schools, and smoke alarm campaigns.

*However, the existence of a program does not ensure that it has been tested and is effective with its intended audience. Local decision makers must therefore be cautious when choosing a program to adapt for their own use.* More is covered in Chapter 12 on documenting impacts and outcomes of prevention programs. But it is important to note that in order to be considered as a "best practice," targeted programs must demonstrate that they were in fact designed appropriately and have produced some measurable results.

As previously indicated, the *Learn Not to Burn* curriculum produced by NFPA still represents a comprehensive approach to the many fire safety problems that exist, and it can be used as a foundation for targeted programs. The topics and material are still appropriate. Many successful targeted programs have been developed using such educational messages as a base—where the research for age-appropriate material has already been done. The *Risk Watch* curriculum is still an example of the more frequently encountered injuries a modern fire department faces and provides methodology for dealing with those specific problems, though it is not actively supported by NFPA at present. More recently, the Home Safety Council has produced specialized educational programs designed to reach audiences who are functionally illiterate and preschool children. They also maintain a specific Web site that holds an interactive program where common home hazards—and accompanying safety messages—may be explored by anyone with access to the Internet.

These types of targeted programs have been well researched, and evaluations have been done to document that learning actually takes place. Behavior changes among target audiences have also been demonstrated.

## SEASONAL AND SPECIALIZED PROGRAMS

Some targeted programs are related less to specific audiences than to a time of year. Consequently, many fire departments have designed programs to deal with problems such as Christmas tree fire safety and the use of fireworks around the Fourth of July. Although there is a coalition of fire service organizations working to eliminate the use of consumer fireworks, they are still legal in most U.S. states, and many local jurisdictions have aggressive public education campaigns (accompanied by aggressive enforcement of time curfews and controlling illegal fireworks) around the Fourth of July and, in some cases, New Year's Eve. Many believe in any event that they have an obligation to teach safe handling practices for consumer fireworks when they are allowed.

## JUVENILE FIRESETTERS

Specific programs have also been developed to deal with juvenile firesetting behaviors. The USFA has developed several manuals that guide local fire officials in a field that ultimately involves the mental health community as a critical partner in reducing these types of incidents.

The USFA has developed and published a manual to help fire departments determine the proper treatment of juvenile firesetters. It helps to determine if children who set fires fall into categories of normal curiosity or are acting out some emotional disturbance. Prepared by Jessica Gaynor, Ph.D., for SocioTechnical Research Applications, Inc., the title of the manual is *Juvenile Firesetter Intervention Handbook*.[8] It may be obtained from the Federal Emergency Management Agency, USFA, in Washington, DC.

Information on some nationally recognized experts in the field of juvenile firesetting can be found through a search at the U.S. Fire Administration. People such as Pat Mieszala of Burn Concerns, Judy Okulitch of the Oregon State Fire Marshal's Office, Don Porth of Portland Fire & Rescue, and Paul Schwartzman of Fairport Counseling have been putting forth heart and effort for many years and can be found through the U.S. Fire Administration Web page or staff.

Specially packaged targeted educational programs should be designed or purchased for the intended problem and audience. Often, the materials are already produced, tested to provide evidence of their effectiveness, and available at a reasonable cost. Local officials responsible for the management of public education programs should review the testing and validation procedures of any product, whether locally produced or taken from national efforts, before borrowing a program or idea from another source, or producing their own. If materials are not available to suit local needs, then local design and production may be inevitable. When designing targeted educational programs, an analysis of the local loss data and a deliberate planning process should be used to prioritize the efforts where they will accomplish the most.

Public education programs are not limited solely to these two broad areas. In fact, there is a great deal of overlap between this method of prevention and the other methods used in a comprehensive program.

## COMBINED PREVENTION PROGRAMS

**combined prevention programs**
■ the partnering of various techniques, such as education and enforcement—or technology and education—for best results

**Combined prevention programs** are those that partner various techniques, such as education and enforcement—or technology and education—for best results.

Not all educational messages are purely educational. They contain elements of other prevention strategies. There are many opportunities for fire safety educators to use the education principles discussed in this chapter in other related ways. For example, the most often promoted engineering solutions to the fire problem are smoke alarms and residential fire sprinklers. Although they are among the best ways to mitigate the effects of fire, they are not always accepted as a cost-effective measure. Getting these fire protection features installed and maintained may require the use of the same educational principles that serve as the foundation of other targeted efforts.

There are a number of educational programs designed to promote residential fire sprinklers. The National Home Fire Sprinkler Coalition serves as a focal point for educational efforts designed to promote the use of sprinklers. The National Fire Sprinkler Association and the American Fire Sprinkler Association also promote the use of education to remove barriers to the installation of sprinklers.

In fact, the process used to adopt local codes and ordinances that require smoke alarms or fire sprinklers includes many of the same elements found in pure public education ventures. Identifying target audiences and presenting information to them in a fashion that will change their attitudes about the need for smoke alarms and sprinklers is another example of how public education principles interrelate with engineering and enforcement solutions. It is also common for local officials in charge of code enforcement programs to adopt a model of educating the business community about the need for such regulations.

Public education is, therefore, a critical element to any comprehensive prevention program, because it reaches an audience that is either resistant to or not subject to other methods of controlling losses. In our society, people must still be convinced of the value of an idea before they "buy" it. Engineering solutions and the establishment of codes are no exception to this premise.

Expanding beyond the fire service to look for partners is another way in which the field of public fire and life safety education has evolved. This level of growth represents a strong change in the way public education works to become more effective.

Click Here to tour MySafeHome

The Home Safety Council maintains an online safety home that provides an interactive way for audiences to identify common home hazards and the related safety messages. (*Courtesy of the Home Safety Council*)

# Coalitions and Partnerships

**Coalitions and partnerships** for prevention efforts are becoming a more common occurrence—often bringing more resources than one partner could muster alone. An emerging trend throughout the sphere of government programs is the need to work *with* the community to solve problems. The fire service is currently responding to this trend in some interesting ways in a number of communities throughout North America. The two basic tenets of this movement are the need to establish coalitions of organizations who can work together to solve problems that overlap and the need for local communities to take responsibility for solving their own problems.

All-injury safety school curricula are a good example of how coalitions are created and managed. They are created around the concept that a variety of groups have an interest in general public safety, so the all-injury curriculum actively promotes the establishment of coalitions to help promote its use. It is obvious that the same group of people from a community may represent a high-risk audience for a variety of emergencies. The problems that arise from crime, drug use, fires, and other societal concerns are usually more prevalent in low-income neighborhoods. Put more simply, it is the same people who must receive the attention of those trying to prevent crime, fires, or other emergency incidents. Consequently, the coalition that supports a program such as all-injury prevention in the schools can also be expanded to deliver the same or similar messages to the community at large.

coalitions and partnerships
▪ groups or organizations coming together to provide more resources than one partner could muster alone

Many departments are utilizing these types of coalitions to help promote their safety efforts in a variety of ways. An often overlooked portion of the five-step planning process is the call for citizen involvement in the problem-solving portion of the plan. The Phoenix Fire Department has used a steering committee of citizens to help promote fire safety for many years. The Philadelphia Fire Department in Pennsylvania used community members and business partners to help promote their smoke detector giveaway program. The Oklahoma City Fire Department used similar techniques to develop their own version of a smoke detector giveaway program that achieved significant results.

The underlying principle in working with coalitions is the premise that communities must take an active interest in and responsibility for solving their own problems. In doing so, they often identify resources and problem-solving techniques not normally available to the fire department. For example, the Houston, Texas, Cease Fire coalition raised more than $500,000 to help with public fire safety programs in their community.

It is important to note that the use of coalitions is increasing in other parts of our society. Mothers Against Drunk Driving (MADD) is an excellent example of a large group of people who come together for a common cause. The success of their efforts at educating the national community, and even getting laws changed, has been noteworthy. Few have missed their efforts or would fail to recognize their impact. The fire service must also be prepared to promote and work with coalitions to further their own safety objectives.

# Arson Prevention and Other Combined Educational Programs

Arson prevention programs are unique among public education efforts. Because deep-rooted psychological problems may play a role in arson, traditional educational approaches may not be effective. Identifying and dealing with disturbed children or adults who experiment with fire, or deliberately cause fires, is more complicated than creating an age-appropriate message and delivering it. Professional analysis and counseling will be needed to change attitudes and behaviors. And if profit is a motivator for setting fires, no amount of education will help. A strong investigation and prosecution effort will be needed to identify, arrest, and convict arsonists motivated by profit.

Many jurisdictions have found that a natural partnership with insurance companies is a good way to obtain resources for arson prevention efforts. Effective arson prevention programs sometimes include an aggressive investigation and prosecution team. They also include property educational components to identify potential targets and prepare them to minimize random or deliberate arson attempts. Sometimes there is even a link with transient populations who are homeless and unintentionally or deliberately set fires.

In any case, there is a link between public education programs and a variety of other methodologies we use to control the fire problem in our communities. The strongest are those that incorporate education, awareness, enforcement, and even technological changes to provide for public safety. These combined efforts utilize elements of other prevention strategies in combination with aggressive education to maximize the impact and outcomes of programs and efforts.

# Summary

Fire officials and other local decision makers must understand that public education and public information or relations may relate to one another, but they are separate functions of a comprehensive prevention program. Information campaigns raise awareness levels, whereas education raises the cognitive level of understanding and changes attitudes about a safety topic to cause a behavior change. A change in behavior can lead to reduced risk within a community and ultimately will be measured by its ability to help reduce or control losses. Educators should understand that effective public education programs are planned and that there are several basic parts of the planning process used to help guide efforts toward their goal. They should know that school programs represent a foundation of educational activities, but that they must be properly designed and tested to be valid. Educators should also recognize that school programs will not cover everyone in a community and that additional targeted programs may be used to attack real or potential problems with positive effects.

Those planning for effective public education strategies should recognize the overlap that exists between education and the other prevention strategies of engineering and enforcement. They should know that one of the most effective ways to have an impact on the wide variety of emergency problems facing a community is to get the community involved in solving its own problems. Forming coalitions that include other organizations is often the most effective way to gather support for these vital public safety concerns.

# Case Study

### Designing an Effective Public Education Program

The Everywhere Community is a suburb of a major metropolitan area. With 92 square miles of territory and a population of just over 200,000 people, it is a growing community with an increasingly diverse population. An influx of Asian, Russian-speaking, and Hispanic people is changing the demographics of the community. Low-income areas of the community are known, but the resources available to gather and analyze data are very limited. It is possible to identify where fires are occurring, but matching that information to specific audiences is almost impossible. The major causes of incidents are known, but not in detail. Of the calls received for emergency response, 80% are medical emergencies. The second highest single call category for assistance is for falls. The specific nature of causes for fires is impossible to obtain except by hand-tabulating more than 20,000 call records.

The fire department offers a fully comprehensive fire prevention program that includes fire investigation, plan review and inspections for compliance for new construction, regular fire inspections for code compliance, and a public education program. The public education efforts are coordinated by a single employee. Additional resources are provided when they are available in the form of other prevention personnel or from on-duty firefighters. Financial resources are limited and are being examined for possible cuts.

The challenge faced by the community and the department is to defend against budget cuts for public education programs. A decision must be made about maintaining a first-grade safety education program that requires significant overtime to visit every first-grade student in the community. Local decision makers are questioning whether or not to continue that program or replace it with more specific public education efforts.

## Case Study Review Questions

1. What is the difference between education and information? How do the two areas relate to one another? Where does overlap exist?
2. List the steps of the planning process for public fire and life safety educational programs and describe what each of them covers. Describe how implementation and evaluation are dealt with *during* the planning process.
3. Describe the critical elements of a school program in terms of educational methodology. Discuss why it is so important to the success of the program.
4. Describe the advantages of a school-based safety curriculum and what factors must be considered before developing or purchasing one.
5. Given the Case Study material, what program elements could be strengthened based on the overall fire and injury problem? What role would the collection and analysis of data play?
6. What suggestions would you make about the use of coalitions in everywhere or another community facing problems with specific target populations?
7. What is the important thing to keep in mind about fire department–based juvenile firesetter programs?
8. What are combined prevention programs?
9. Do public education programs actually work?
10. Does installing smoke alarms constitute an educational program?

PEARSON
### myfirekit™

For additional review and practice tests, visit **www.bradybooks.com** and click on MyBradyKit to access book-specific resources for this text! Your instructor may also assign Additional Project work related to topics in this chapter.

Register your access code from the front of our book by going to **www.bradybooks.com** and selecting the mykit links. If the code has already been scratched off, go to **www.bradybooks.com** and follow the MyBradyKit link from there.

## Endnotes

1. Bachtler, Joseph R., and Thomas F. Brennan, *The Fire Chief's Handbook*. Saddle Brook, NJ: Fire Engineering Books and Videos.

2. *Proving Public Fire Education Works.* Arlington, VA: TriData Corp., 1990.

3. *The Community-Based Fire Safety Education Handbook.* Washington, DC: Rossomando and Associates for the National Association of State Fire Marshals, 1996.

4. *Reaching High-Risk Groups: The Community-Based Fire Safety Program.* Washington, DC: Rossomando and Associates for the National Association of State Fire Marshals, 1996.

5. *Public Fire Education,* 5th ed. Stillwater, OK: International Fire Service Training Association, 1987.

6. The *Learn Not to Burn* curriculum was developed by the National Fire Protection Association in 1998.

7. *Risk Watch,* an all-injury safety curriculum, was developed by the National Fire Protection Association in 1998.

8. Gaynor, Jessica, *Juvenile Firesetter Intervention Handbook.* Washington, DC: U.S. Fire Administration, Federal Emergency Management Agency, 2000.

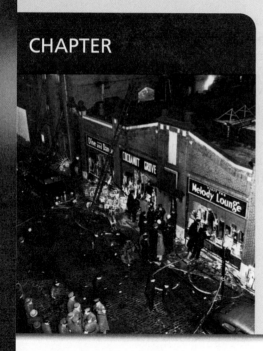

# 3

# Concepts in Code Enforcement

3.1 National Commission on Fire Prevention
Manufacturer's Association, ISO, Insurance and Association for Fire Service Instruction
3.2 Smith (1979) — Adopt a...
3.2.1 Fire History
Laws, Five-Foot Clear...
3.3 Interpretations and Revisions...
System, Interpret a Code History...
3.4 Federal, State, and Local Interest — Administrative
International ICC, NFPA Standards, UStat...

**Appeal process** *p. 55*

**Code administration** *p. 55*

**Code development** *p. 53*

**Fire protection contractors** *p. 66*

**Inspection process** *p. 61*

**Occupancies** *p. 57*

**Operational permits** *p. 61*

## OBJECTIVES

After reading this chapter you should be able to:

- Point out major historical occurrences in code development.
- Contrast codes and standards.
- Describe the legal basis for code enforcement and the interfaces among national, state, and local codes.
- Describe elements of code administration, inspection practices, and the appeal process.
- Differentiate among code interpretation, revision, and retroactivity.
- Describe the mini-max concept of code enforcement and list a variety of code promulgation organizations and ramifications stemming from the globalization of codes and standards.

# Developing and Administering Quality Codes and Making Sure They Are Enforced

Code enforcement is traditionally thought of as a principal portion of a comprehensive fire prevention program.

Many fire departments put most of their prevention resources into code enforcement activities. Some do so with designated (full-time) fire prevention personnel, and some do so by using emergency response crews to conduct inspections. Some departments do not perform code enforcement activities at all—relegating them to another government agency, or leaving them entirely undone. However the function is managed (or not managed), it is an important part of a fire department's arsenal in combating fire losses.

## HISTORY OF CODE DEVELOPMENT

History has taught us that the fire codes originated from many disasters in the past. As previously mentioned in Chapter 1, there have been many fire disasters throughout history that have resulted in the formation of one or more portions of more modern fire codes and standards. Among them are the 1911 fire in the Asch building in New York City. Commonly referred to as the Triangle Shirtwaist fire, it resulted in the deaths of 146 young women who worked in the garment factory. Many died jumping out of windows trying to avoid being killed by the fire. Most credit that fire and ensuing efforts to promote more effective prevention efforts as principal factors in the establishment of the *Life Safety Code* (NFPA 101) produced by the National Fire Protection Association.

The Cocoanut Grove nightclub fire in Boston, Massachusetts, occurred in 1942, killing 492 people in an overcrowded club with limited exits and combustible walls and decorations.

Other examples include the Beverly Hills Supper Club fire in Southgate, Kentucky. This fire occurred in the now-infamous nightclub on May 28, 1977, and killed 165 people. Another is the MGM Grand Hotel Fire in Clark County, Nevada. That fire, which occurred in 1980, killed 85 people and injured another 600.

More recently, the terrorist attacks on the World Trade Center towers in New York (September 11, 2001) demonstrated that thousands could die in a modern high-rise building—despite previous codes that provided safety features, including fire protection for steel structural elements. No one would have considered the possibility of an airplane (let alone two) flying into a high-rise building in our nation. Most buildings are not designed to withstand such an external blow and

The Cocoanut Grove nightclub fire, which occurred in 1942, was one of the worst fire disasters in U.S. history. It helped pave the way for many new code requirements, especially adequate exiting in public establishments. (*From the Collection of William Noonan, Boston FD*)

the ensuing fire, which is also fueled heavily by the jet fuel inside the airplane. But even a neighboring building eventually collapsed from fire—and it had not had a jet crash into it. The collapse of the buildings at the World Trade Center, caused ultimately by fires, provided lessons about adequate exiting and protection of steel structural supports.

These fires, and others throughout history, have taught us lessons about fire safety that have been incorporated into modern fire and building codes. Laws requiring smoke alarms and codes that require adequate exiting for emergencies and fire sprinklers stem from these lessons. Code requirements for specific hazards are usually linked back to specific fire problems in our history, though some were put in place because of potential risks involved and anticipated safety measures that could prevent such disasters.

However, fire and building codes are developed, any comprehensive prevention effort must include mechanisms to ensure that the codes are followed.

## CODES VERSUS STANDARDS

Laws and codes developed with the best of intentions accomplish nothing if they are not followed. Most people will comply with laws and codes if they understand the need for doing so, but an active enforcement program is necessary to ensure that everyone will. An unfortunate feature of modern society is that not everyone sees the same need for fire safety codes. Frequently, businesses are far more concerned with their financial interests than with the safety of their operation. However, the codes were not developed to protect only the proprietor. They were also developed to protect the lives of customers and of the firefighters who must respond if a fire occurs.

But fire codes do not stand alone. A basic building block for codes is the set of standards used as their foundation. Codes generally outline what is required, whereas standards stipulate how the requirement is to be met. For example, a fire or building code would require fire sprinklers in certain circumstances. However, requiring them is not enough. Outlining the specific installation and hydraulic standards necessary to make sure they will function properly is the role standards play in the **code development** and enforcement process. Those installation standards would be found in NFPA 13—the standard that deals specifically with fire sprinkler systems. This interrelationship is also discussed in Chapter 4, which deals with the plan review and acceptance inspection process. Codes *and* standards are necessary for a comprehensive fire and life safety prevention program. Among the more important standards promulgated by the National Fire Protection Association are NFPA 72—the standard that deals with smoke detectors and smoke alarm systems (for commercial occupancies) and smoke alarms for residential use.

code development
■ process of creating code provisions

Another aspect of this relationship is demonstrated in other standards. The code requires fire sprinklers for certain occupancy types. But how they are to be inspected, tested, and maintained is found in yet another related standard (NFPA 25). Another example of this relationship can be found in NFPA 96, the standard that deals with inspection, testing, and maintenance of kitchen venting and fire protection features.

These standards can be critical for fire protection, and a great deal of effort has been expended on them to focus on particular problems. For example, in recent years there has been a tremendous amount of energy devoted to determining the best type of smoke alarm for residential settings. It is widely held that ionization-type smoke alarms are more effective (alert more quickly) in fast-spreading fires. Conversely, photoelectric smoke alarms alert more quickly to slow, smoldering fires. There is no general consensus at present on which type works "best," and study continues to see if an answer such as dual-chamber (ionization and photoelectric) alarms might provide the best overall protection.

It is important to note that the place where these discussions occur is usually in the standards development process. Once these issues are decided, the fire code would deal most frequently with the issues of which occupancies would be required to have smoke alarms or smoke detection systems.

There are numerous standards that provide much more detail than the fire codes about how systems should be designed, installed, tested, and maintained. They form a foundation for code development.

There are several other fundamental issues about code enforcement that must be understood by fire officials and local decision makers. Among them are the authority to adopt and administer codes, the management of an appeal process, the development of codes and laws, and the globalization of codes and standards. First among the issues of concern for local jurisdictions is the authority from which the ability to enforce laws is derived.

# Authority of the Code Administration Process

Generally, the authority to administer codes, laws, and their underlying standards is passed down from the state and establishes what is commonly called the *authority having jurisdiction.* This term refers to the official who has the responsibility for managing and enforcing the codes and laws adopted by that particular agency and within the scope of their jurisdiction. Such an official is commonly referred to as the AHJ.

The codes and laws relating to fire safety, which are adopted by local jurisdictions and individual states, may usually be found as part of "model" fire codes or specific state statutes. There are exceptions to this, because some jurisdictions develop and adopt their own construction and fire safety codes. Generally, this is to ensure a large degree of local control over code requirements. Until recently, New York City had its own codes, but moved to adopt nationally recognized model codes as their foundation. Many jurisdictions are finding that developing codes at the local level may provide more control, but also incurs some risk because the resulting codes may not match the national consensus on code or standard requirements, potentially increasing local liability.

Model codes are those produced regionally, nationally, or internationally to serve as a foundation of modern code enforcement concepts and language. In addition, some states establish laws relating to fire reporting, smoke detectors, fire sprinklers, or fireworks. Many localities adopt a statewide version of a model fire code, but model fire code language is useless to a local jurisdiction unless it is legally adopted for that community. Fire officials and other decision makers must understand which responsibilities belong to the federal government, the state, and the local jurisdiction. Many times, the distinction is not clear, and occasionally concerns about duplicating enforcement services arise.

For example, some states maintain enough staffing to conduct field inspections of certain occupancies such as hospitals. Because hospitals must meet national medical accreditation standards, they are often held to safety standards outlined by NFPA 101, the Life Safety Code. Concurrently, many local jurisdictions may be conducting fire code compliance inspections of their own, operating most often (in the United States) under the International Fire Code, though a significant number use NFPA 1 as their model fire code. Fire code enforcement inspections conducted by two different agencies, using two different codes, can be alarming to local administrators. Such a situation can raise questions of ultimate authority for jurisdiction for code enforcement. In most of the codes, the term *authority having jurisdiction* means that the authority for actually enforcing the provisions of the code has been clearly identified and resides with a specific agency and individual. But conflict can and does arise. It can also arise from overlap between various codes. Those jurisdictional issues are discussed further in Chapter 4.

Most model codes have been in place for many years, and their underlying authority is taken for granted. But legal challenges may arise, and, especially for newly formed fire departments, the foundation of legal authority may not be clear. Therefore, it is most important to obtain proper legal advice locally to understand both the legal authority and limits of a jurisdiction's code enforcement efforts.

Adopting codes and laws is only part of the code enforcement process. Administering them is often an art that requires sound judgment and open communication between those responsible for enforcement and those who must comply.

# Administering Codes

Enforcing codes refers to the individual responsible for gaining compliance in the field. Those who do the day-to-day inspections are enforcing the fire code. *Administering code enforcement activities* refers to something more complex and is not an easy task. Administering code enforcement requires a great deal of oversight, because many aspects of a fire code are complex and often subject to interpretation.

Because the codes are complex and understanding takes a great deal of time and training, individual inspectors have many opportunities to vary from prescribed performance. It is, for a fire marshal (usually the head of code enforcement activities) to hear complaints about inconsistent code application. For example, one inspector enforces one provision of the code; another inspector may miss it and enforce another provision. This issue can be exacerbated by the quality of training and experience of varying inspectors. In addition, it can often be used as a manipulative ploy from some businesses to try to get around code provisions. Some business owners may attempt to deceive the local authority by implying that other jurisdictions interpret the code differently, or that a previous inspection did not enforce a provision so they should not be required to adhere to the code as it presently is stated. Sometimes this is true—and other times it is not.

Consistent application of the codes in an enforcement setting is always difficult. Usually, legal representatives will stipulate that under the best of circumstances, some provisions of the code may be missed. But once missed provisions are identified, they must be adhered to, or alternate means of compliance must be found. Therefore, even if missed provisions are found late in the inspection process, or found by another inspector, code provisions must be met or varied from in a specific fashion.

## MODIFYING CODE REQUIREMENTS: THE ADMINISTRATIVE APPEAL PROCESS

The administrative **appeal process** in this context refers to the ability of an AHJ to modify the code requirements without a formal board of appeals. The AHJ does so either by accepting alternate ways of accomplishing objectives, or by waiving requirements that are found to be impractical. Some risk is involved because liability can increase. However, this part of **code administration** is valuable because it limits liability for individual inspectors, and it defers decisions about code alteration to a higher authority where due diligence about risks and code alternatives may be more adequately carried out.

**appeal process**
■ outside board of citizens within the authority having jurisdiction

**code administration**
■ managing the code enforcement process

There are provisions in the model codes for variances from the pure requirements prescribed in them. Public officials often hear concerns from constituents about the cost of safety requirements when they are forced to comply with the codes. Some may want provisions waived solely on the basis of economics. However, any authority ignoring the provisions of the code does so at great peril. Ignoring the code increases the likelihood that the authority's agency could be included in liability battles over losses, should a fire occur. Some latitude may be given on the time required to comply with code provisions. At other times alternative materials and methods may provide a more cost-effective mechanism to produce public safety. But even when an inexperienced inspector has missed a code provision, the codes must be enforced where resources exist to do so.

In general, when conducting inspections, enforcing the codes by following exact model code language is the simplest method of achieving safety and consistency. It also keeps the AHJ from becoming involved in every inspection activity or decision.

As previously mentioned, not all fire departments have been given the authority to administer their code enforcement activities. For those that do administer

# VANCOUVER FIRE DEPARTMENT

## Fire Marshal's Office Policy

| Policy #:<br>4.005 | Subject of Policy:<br>Fire Sprinkler System Flow Alarms | | |
|---|---|---|---|
| Developed By:<br>Chad Lawry | Title:<br>Deputy Fire Marshal | | Date:<br>12/16/03 |
| Reviewed and Approved By:<br><br>**Jim Crawford** | Title:<br><br>Fire Marshal | | Effective Date:<br>12/16/03<br>Next Review:<br>04/20/2008 |

THIS POLICY IS TO SERVE AS A GUIDE FOR PLAN REVIEW AND ACCEPTANCE TESTING. EXCEPTIONS TO THIS POLICY SHALL BE APPROVED BY THE FIRE MARSHAL THROUGH THE APPROPRIATE SUPERVISOR.

**PURPOSE:** This decision applies to development of policies and procedures, inspection services and interpretations of the Fire Code to the Plan Review and Permit process administered by the City of Vancouver office for Development Review Services.

**GENERAL:**

1) Flow Alarms are required on all Fire Sprinkler Systems.
   - One audible flow alarm shall be provided on the exterior of the building in an approved location.
   - One audible flow alarm shall be provided in the interior of a building in a normally occupied location.

2) Flow Alarms in Multiple Tenant Occupancies.
   - Multiple flow alarms are not required at buildings that are served by one fire sprinkler system and service.
   - Where separate fire sprinkler systems/services are installed for each tenant space or a portion(s) of a building, each system will be provided with an exterior and interior audible flow alarm.

3) Actuation of Audible Flow Alarm.
   - Actuation of the audible flow alarm shall be as set forth NFPA 13 and NFPA 72.

The Building Official and the Fire Marshal are vested with the power to render interpretations of the code and to adopt and enforce rules and regulations to supplement the code as deemed necessary to clarify the application of the provisions of the code.

code enforcement activities, policies are usually required to help guide the variety of personnel who are responsible for enforcement at different levels. There is a difference between law and policy for code officials. Laws and codes usually state the prescriptive requirements (e.g., where smoke alarms are required), whereas policies (much like standards) deal with the more specific "how to" questions that arise during the daily administration of codes. For example, a law may state that a smoke alarm is required in every home; the standard may specify where the alarm should be placed and what type of alarm is appropriate. The policy would then describe how that law would be enforced. More specifically, a policy might state that enforcers would gain compliance by issuing a specific form. If inadequate resources exist to ensure compliance annually, a policy might also address the frequency of inspections or how a self-inspection by building occupants might occur.

In addition, many codes require some interpretation as to their exact meaning. Some codes leave decisions about specific issues up to the local fire chief: for example, the chief may determine what constitutes an imminent fire hazard requiring drastic enforcement measures. Such authority is very broad in theory, but rarely used in practice because of practical political concerns. Fire administrators who think they are the final word in code enforcement may be naïve about the pressures that can be brought to bear by concerned citizens who perceive an abuse of power by fire officials or decision makers. Under these circumstances, clear policies establishing code administration practices can provide the consistency desired by the community and offer guidelines that will help enforcers administer codes in an equitable fashion. Whenever discretion is called for, it is wise to make sure that political leaders will support the conclusions drawn by those directly responsible for code enforcement personnel.

For example, it is still largely a matter of policy, not law, as to how often code compliance inspections must be done. A local authority that inspects a specific business more frequently than others may be subject to complaints about harassment. Unless a more aggressive inspection cycle is warranted by the hazards the business occupancy presents, decision makers could agree with a particular business owner about the practices of their own local code enforcement official. Policies may or may not have the same force as law, and again, local jurisdictions should obtain legal advice from their jurisdiction's counsel about their ability to enforce policies as well as law. But policies can be critical and can even result in job changes for those deemed to be too strict in their interpretation of the administration process.

There are some resources that outline the steps of administering the code enforcement process. Two such resources are the *Fire Inspection Management Guidelines*[1] and *Conducting Fire Inspection*.[2] In addition, the International Fire Service Training Association book *Fire Inspection and Code Enforcement*[3] is an excellent resource on the topic of code enforcement and inspections. These books discuss some issues of concern for managing a code enforcement program, such as identifying an inventory of properties to be inspected, maintaining a database and record-keeping system, inspection procedures, selecting and training personnel, and managing appeals. Each of these areas of concern is addressed in the following sections.

## INVENTORY OF PROPERTIES AND INSPECTION FREQUENCY

The goal of a code enforcement inspection program is to ensure that regulated properties comply with appropriate fire safety codes, though inspections also provide an opportunity to make nonbinding fire and life safety recommendations. Business properties are commonly called **occupancies**, referring to the individual business operation at a particular location.

**occupancies**
■ distinct business operations. A building might contain one or many occupancies, each with a different operation and sometimes different occupancy classifications.

That could be an office within a building, an entire building, or sometimes leased space inside one particular room. The business occupancy refers to a distinct operation for which the responsible party would be different and distinct. There are also different occupancy types, each with its own set of specific code requirements. For example, a public assembly occupancy, such as a nightclub, would have very different exiting requirements from a business occupancy, such as a small convenience store. The model codes outline specific occupancy types and accompanying requirements.

Some fire departments have adopted a philosophy that uses enforcement actions only after their inspectors have attempted to gain willing compliance. In many respects, the code enforcement and inspection process is an opportunity to educate the community about proper safety measures. In the past, more stringent measures have been used to enforce provision of the codes, but only when the public clearly understands the nature of the hazard. That situation does exist in other parts of the world. According to TriData studies Japanese society, because of the close proximity of housing units to each other, has a much greater emphasis on fire safety.[4] It is a matter of public shame to have a fire: families are sometimes forced to relocate because of the ire of their neighbors over an unintentional fire. Under those circumstances, severe penalties for code violations would be less likely to cause concern among the public or political decision makers.

But in the United States, where a general culture exists that limits government interference, stringent enforcement penalties are often viewed as draconian. That is why issues such as inspection frequency and the philosophy behind enforcement can be critical elements in the administration of the code compliance process.

Ideally, every commercial property in a jurisdiction would receive regular, systematic fire safety inspections. However, this is not a normal practice for many jurisdictions. When first identifying an inventory of properties to be inspected, local decision makers may be forced to prioritize their inspections based on a statistical history of fire problems or on a list of potential problem occupancies where the risk of death or loss is great, if infrequent. An example of potential problem occupancies would include hospitals or schools, where the number of true fire incidents is usually small, but the potential for disaster is great if they do occur. Consequently, prioritized inspection practices may include a combination of those properties where fires are frequent and those where the risk is great.

The frequency of inspections is predicated on the resources available for the task. Obviously, the relationship between prioritizing inspections and the resources available will yield some kind of inspection cycle that is appropriate based on a risk assessment. Generally speaking, more frequent inspections would lead to a safer community. If nothing else, a local jurisdiction would be able to document the number of hazards noted and abated through the inspection process, thus providing some evidence that risk was being reduced. However, no definitive studies have outlined what inspection frequency is best. One study, performed by the National Fire Protection Association, did identify a relationship between inspection frequency and the number of fires.[5] It did not produce a specific time frame that should be applied for all jurisdictions and all property types. More recent studies, not yet completed, also raise questions about which types of inspections are more effective at managing risk and eliminating potential hazards. These studies indicate that in many cases, the philosophy behind inspections (such as educating business owners about hazards) may be every bit as important as the inspection cycle. Readers can learn more about these studies by going to NFPA's Web page and searching under the fire research foundation reports. Until better data is available, local decision makers have no choice but to outline the best (and most frequent) inspection cycle they can effectively manage with the resources allocated for the task.

# MAINTAINING A DATABASE AND RECORD-KEEPING SYSTEM

Maintaining a database and record-keeping system is an important aid in prioritizing inspections. Such a system can help determine where fires are occurring and how often buildings should be inspected. When a fire occurs, decision makers usually

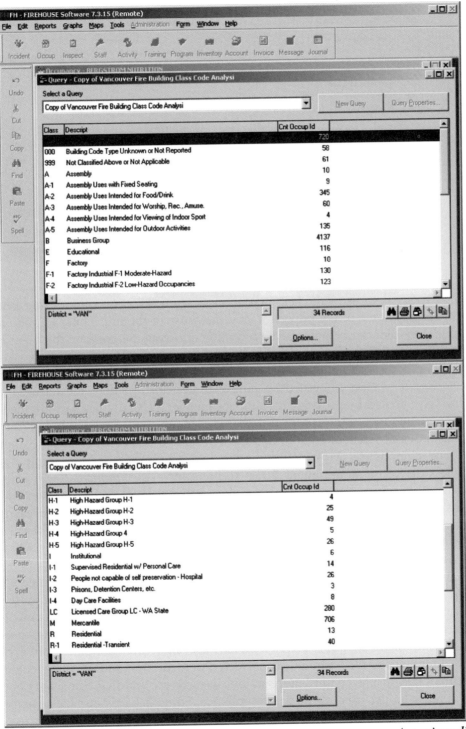

A database of inspectable occupancies may be a simple file, or a complex computer program such as those offered commercially. These programs may be readily found with an Internet search or at trade shows. (*Courtesy of Vancouver Fire Department, WA*)

(*continued*)

(continued)

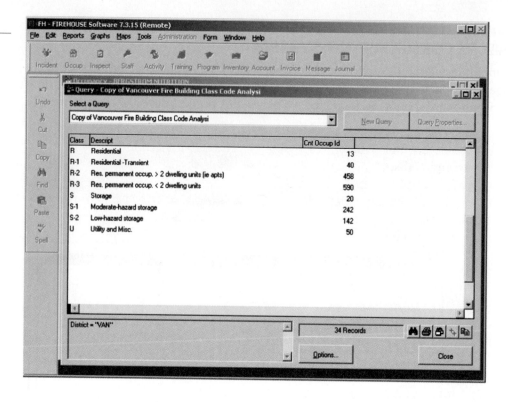

want to see records of the last inspection to see if there is a relationship between the hazards noted in the inspection and the cause of the fire. Thus, a good record-keeping system can help improve prevention efforts by pinpointing the hazards and business types that most commonly lead to fires. A good record-keeping system is also important for legal purposes, as discussed later in the chapter.

Another aspect of record keeping includes documenting the hazards found and later abated by property owners or managers. Usually hazards are identified through an inspection process and abated through some type of administrative or quasijudicial system of code compliance. In some cases, court orders are sought to obtain compliance. Most jurisdictions allow 30 days to correct simple hazards unless they pose an imminent threat to life safety. An *imminent threat* is loosely defined in most codes and requires judgment from local decision makers. It usually indicates a breakdown in a fundamental fire safety feature that threatens the safety of occupants. Typical examples include locked exits, disabled fire sprinkler or alarm systems, and placement of people in a hazardous area for which the building was not designed. Finding people living in a building that was not designed for their safe exit is considered a threat by fire officials, who are concerned for the safety of residents and for their own liability if they allow an unsafe practice to continue unabated. People living and sleeping in the wrong types of buildings also represent a significant rescue challenge for firefighters. In these cases, the potential for the loss of multiple lives exists, even before firefighters can arrive.

Imminent hazards should be corrected immediately, and at times tenants are temporarily displaced until more serious hazards are rectified. Some may be given a slightly longer time frame for compliance. Some hazards involve structural or mechanical problems, which may require longer periods of time to correct. The variety of problems encountered makes it impossible to specify a "one size fits all" approach to hazard abatement. It is up to the local authority to determine an acceptable timetable for correction of hazards and to ensure that there is political support for these decisions. Also, for liability purposes, many code compliance forms stipulate an immediate correction time frame for all corrections, but allow a window of time between the initial inspection that identified the hazard and the reinspection that will occur to make sure it was abated. This means that the fire hazard is not being allowed to continue, and that the owner is responsible for fixing it immediately. But it also gives an owner or operator *time* in the real world to comply without absolving him or her of the responsibility to maintain a safe building that complies with the fire code.

All of the elements of the inspection process must be documented to ensure that the public interest is properly managed. Liability is another issue that is discussed later in the chapter.

## THE INSPECTION PROCESS

The **inspection process** represents the procedures the individual code enforcement personnel would follow to conduct an inspection. This chapter refers to the inspection process associated with regular fire code enforcement activities, as opposed to those that are associated specifically with new construction, permitting, acceptance inspections, and issuing certificates of occupancy.

**inspection process**
■ procedures followed by code enforcement personnel to conduct an inspection

The inspection process in this context refers to the process of identifying occupancies to be inspected, researching past inspection records, conducting inspections and identifying hazards, conducting follow-up inspections to ensure that hazards are corrected, and the maintenance of inspection records to document that hazards have been identified and corrected.

There are some excellent texts that deal with the inspection process and its complexities in far more detail than this chapter. Among them are manuals produced by the International Code Council and the National Fire Protection Association. The International Fire Service Training Association also produces training manuals that deal in depth with the inspection process. But managers should understand the basics of a fire code compliance inspection and how it fits into the larger code administration process.

The individual inspection process begins with an understanding of which **operational permits** are required by the code.

Operational permits are part of the every day use of a building rather than a particular construction requirement or feature. An example would be an operational permit required for open flames or torches—allowing the fire code official to monitor and enforce provisions of the code related to use of torches for paint removal. This could occur in any occupancy type—regardless of its approved use during construction—hence requiring an operational permit on the occasions when it might be necessary.

**operational permits**
■ required to outline storage and use requirements for specific hazards identified in the code

The next step is generally to identify the occupancy type to be inspected. As previously mentioned, an assembly occupancy would usually have different operational permits and different code requirements than another occupancy type such as a mercantile occupancy. Each might have different requirements for housekeeping, electrical safety, hazardous materials, and fire protection features. So understanding the code requirements for a specific occupancy type is necessary and is an identified step in the inspection process.

## PALM BEACH GARDENS FIRE DEPARTMENT
10500 N. Military Trail, Palm Beach Gardens, FL 33410  Phone (561) 775-8260  Fax (561) 775-8269

### FIRE SAFETY INSPECTION REPORT

Occupant: _____  Date: _____

Address: _____  Suite: _____  Phone No.: _____

Emergency Contact: _____  Title: _____  Phone No.: _____

| | |
|---|---|
| ( ) | **Means of Egress:** ( ) obstructed  ( ) locked  ( ) improper locking device<br>Location(s): |
| ( ) | **Emergency Lighting:** ( ) inoperative  ( ) not provided<br>Location(s): |
| ( ) | **Illuminated Exit Sign:** ( ) inoperative  ( ) not provided  ( ) battery back-up inoperative<br>Location(s): |
| ( ) | **Fire Sprinkler System:** ( ) required clearance (18 inches) not maintained for fire sprinkler heads<br>( ) not monitored by fire alarm system  ( ) no maintenance contract *NFPA 25*<br>**Fire Department Connection:** ( ) obstructed  ( ) damaged  ( ) missing caps<br>**Fire Sprinkler Riser:** ( ) valves not chained and/or locked  ( ) spare heads not provided |
| ( ) | **Fire Alarm System:** ( ) nonfunctional  ( ) no current inspection tag  ( ) no central station monitoring |
| ( ) | **Dangerous accumulation of waste or combustible material.**<br>Location(s): |
| ( ) | **Fire Extinguisher(s):** ( ) not provided  ( ) not accessible  ( ) not mounted  ( ) requires maintenance<br>*Minimum classification is 2A-10BC rated extinguisher inspected and tagged by a licensed technician*<br>Location(s): |
| ( ) | **Electrical Hazards:** ( ) improper use of extension cord  ( ) missing blanks in electrical panel<br>( ) use of outlet multiplier  ( ) combustible material too close to a heat producing appliance<br>( ) electrical panel obstructed — *minimum 30" clearance*  ( ) circuits not marked<br>Location(s): |
| ( ) | **Address not visible from the roadway:** ( ) front  ( ) rear |
| ( ) | **Knox Box:** ( ) requires maintenance  ( ) missing/improper keys |
| ( ) | |
| ( ) | |
| ( ) | |
| ( ) | |

*Authority:  Chapter 38, Code of Ordinances, City of Palm Beach Gardens*

| | |
|---|---|
| ( ) NO VIOLATIONS NOTED THIS DATE | ( ) FORWARD TO FIRE MARSHAL |
| ( ) ALL VIOLATIONS CORRECTED | ( ) FORWARD TO CODE ENFORCEMENT |

Your immediate attention is required on the above listed items.  You have (14) days to correct any and all violations unless otherwise noted.  Failure to correct violations on or before the re-inspection date will result in further action by the Fire Marshal's office.  Please contact the Fire Prevention Division of the Palm Beach Gardens Fire Department if you have any questions.  (561) 775-8260.

Inspected by: _____  I.D. No. ____, Date: _____  Occupant: _____

Re-inspected by: _____  I.D. No. ____, Date: _____  Occupant: _____

Remarks: _____

White copy – Occupant          Yellow copy – Occupant after violations corrected          Pink copy – Fire Prevention

The next step—highly recommended but often not done—is a review of the previous inspection records. Such a review is not always possible, but it can be critical, for example, to understanding that previous inspections produced some kind of an agreement for a code modification that did not meet model code language. A new inspection conducted without knowledge of such an agreement might directly contradict previous agreements reached. This would cause great concern to property owners who think they are abiding by the law, but are told by a current inspector that they are in violation of the model code language. Without research into previous inspections and code agreements, this information may be missed and result in unnecessary conflict. Reviewing previous inspection records takes additional time, to be sure, but it can save inspection time in the long run. Doing this research before the inspection begins can also point out recurring hazards that may need additional attention because they show up repeatedly.

Generally, the inspection process then follows with a physical visit to the property, during which a thorough inspection is conducted completely around the building and from top to bottom. This process is to identify specific hazards, to note them for correction, and then to physically revisit the property to ensure that the hazards have been abated. Sometimes hazards are so extreme (such as locked exits) that they are required to be corrected immediately. Others are usually ordered to be corrected, and a follow-up inspection scheduled to ensure compliance after a fixed period, commonly 30 days later.

When hazards are not corrected, a process to increase penalties exists to gain compliance within the code. This could involve taking property owners to court or issuing local monetary fines to help gain compliance. It is usually rare to fine or penalize property owners, and often time frames for gaining compliance are negotiated to take into account the reality that some hazard corrections involve considerable expense and time. In some cases, years may be allowed for especially costly code fixes, such as those found when an occupancy use changed for an unwitting business owner—where expensive fire protection systems (such as fire sprinklers) became a requirement because of this new use. As an example, many warehouse owners do not realize that by if they increase their stock size, in-rack fire sprinkler systems might be required. Finding this out during a regular inspection can create a serious financial hardship, however justified it may be by the need for fire safety. In such cases, giving owners time to comply or, as an alternative, forcing them to reduce stock are real-world scenarios that take more time and require additional visits.

Finally, closing out the inspection process means filing proper reports that identify all the hazards found and their correction as a matter of record. In some cases, a written record of agreements for time extensions or alternative solutions with scheduled follow-up for gaining compliance may also be part of the inspection record. One of the worst things a local jurisdiction can do is to identify a fire hazard and leave it unabated.

Some jurisdictions use inspection check sheets such as the one shown earlier to help their inspectors stay on track and gain some measure of consistency. But check sheets are no substitute for proper training and monitoring of inspection activities, with adequate supervision, and in some cases hazards can be missed because they cannot all possibly go onto one inspection checklist.

Inspection procedures may be different for each jurisdiction. However, some common procedures are generally accepted as a standard method for conducting inspections. The first tenet of the inspection process is obtaining permission to conduct an inspection. Local authorities may be able to obtain administrative warrants through the courts to conduct inspections, but the right of entry is not automatically granted to an AHJ. And for states without administrative warrants (such as Washington state), the process for gaining entrance to enforce code compliance can be even more difficult. In that case, a criminal warrant is needed and the burden of evidence for noncompliance is greater to gain entry. This aspect of how to approach inspections is important when selecting and training inspectors, because it implies a voluntary compliance approach first, and enforcement activities second.

To avoid problems, many departments attempt to schedule their inspections, and even to make appointments in advance. This procedure can actually help the overall inspection process by ensuring that a qualified individual is on hand to assist the inspector. It can help to gain willing compliance and ensure that someone on the property is properly educated about hazards so that they do not reoccur.

Some jurisdictions issue preinspection letters before the inspection begins. These letters outline procedures in advance and can also help with the educational aspect of code enforcement. When the actual inspection occurs, it is usually a top-to-bottom physical examination of the business to be inspected. As previously mentioned, this inspection process varies depending on the occupancy type, so preinspection letters can be tailored to specific businesses and their associated hazards.

Preinspection letters help business operators prepare for inspections and clean up hazards before inspectors arrive. They can be a great help with customer relations because they reduce the likelihood of surprise. (*Courtesy of Portland Fire & Rescue, OR*)

Randy Leonard, Commissioner
John Nohr, Fire Marshal
Prevention Division
1300 SE Gideon Street
Portland, OR 97202
(503) 823-3700
Fax (503) 823-3969

Dear Business Owner:

Your business or property is due for a routine fire inspection. Sometime in the near future a Fire Inspector will visit the site and check for fire safety violations. **You do not need to schedule an appointment;** the inspector will drop by when working in the area. If you need to schedule the inspection in advance, call us at 503-823-3700.

Please take a few minutes to review the list of hazards commonly found during inspections, and survey your building for possible problems. We also encourage you to check for any safety issues that are specific to your type of business and correct them immediately. **You will be mailed an invoice for the inspection and violations noted by the inspector**, so you can reduce your cost by eliminating hazards **prior to** the inspection.

Routine fire inspections have proven to reduce fires and associated loss, including death and injury. Our inspectors find and abate thousands of fire hazards each year, some of them quite serious. Since fire damage often affects adjacent buildings and offices, a comprehensive code enforcement program benefits everyone who lives and works in Portland.

For more information about our code enforcement program, including fees, visit our website at www.portlandonline.com/fire/. If you have questions contact us by email at firemarshal@fire.ci.portland.or.us or phone at 503-823-3700.

(*continued*)

## FIRE INSPECTION FORM INFORMATION
### (If further clarification is needed, please contact your District Inspector)

**ITEM #:**

1. **Provide approved address identification for building including suite or unit numbers (letters), plainly visible and contrasting to background color.**
   It is important that your address be visible from the street fronting your property. Suite or Unit numbers/letters should also be plainly visible. For larger complexes or buildings set back from the street, additional signs may be necessary as emergency responders are dispatched to an address. If the address does not readily identify the property location, precious time may be lost time that can save lives.

2. **Remove obstruction and/or combustible material from exit-ways, stairways and fire escapes.**
   The safest way out of your place of business in an emergency is through the exits. Make sure that storage and merchandise is kept clear of exits and exit path.

3. **Secondary exits must be clearly marked. Ensure that illuminated exit signs are fully lit at all times.**
   Generally all exits except the main entrance are required to have an 'EXIT' sign. If there are lighted exit signs, all bulbs must be illuminated when the building is occupied. Replace bulbs and check the operation of emergency lighting to ensure that they will work in the event of an emergency.

4. **Repair fire doors to their automatic self-closing and latching condition. Remove all wedges and doorstops.**
   Doors that have self-closing devices are designed to keep fire and smoke from spreading to other areas of the building. When placing devices that hold these doors in the open position, such as wedges and door-stops, you reduce your chances of escape in the event of a fire. This practice allows for more rapid spread of fire and smoke throughout the building and increases the amount of fire damage.

5. **Discontinue use of extension cords and multi-plug adapters. (UL listed multiple-outlet strips with circuit breakers are acceptable).**
   Extension cords are designed only for use with portable appliances, (i.e.: drill, buffer, grinder) not in place of permanent wiring. Multi-plug adapters are never allowed. The use of extension cords and non-approved, multi-plug adapters may overload the electrical circuit capacity and have been shown to be a major cause of fire. UL Listed multiple-outlet strips with built in circuit breaker protection is allowed in lieu of extension cords. These outlet strips must plug directly into the wall outlet and the appliance must plug into the outlet strip.

6. **Provide cover plates to all open electrical boxes and switches.**
   To confine potential arcing to within the safety of the outlet and junction boxes, replace all missing and damaged cover plates.

7. **Maintain 30" clearance in front of, and clear access to, all electrical panels.**
   Clearance around and access to your electrical panels allows firefighters to quickly access circuit breakers in case of emergency. If your panels are in an enclosed room, provide permanent signage (i.e. "Electrical Room") to indicate location.

8. **Inspect fire extinguisher(s) monthly and maintain written log, or provide annual service.**
   Unless monthly maintenance has been performed by trained individuals and written records maintained, annual maintenance by a certified company is required for portable fire extinguishers. Monthly maintenance checks and written records will, in most cases, allow the service by a certified extinguisher company to be extended to a six (6) year interval.

Whatever the type of building, exits are a critical concern for fire and life safety. These paths of egress are usually designed with heavier construction to help keep fire and smoke from spreading and cutting off escape routes. Buildings are also designed to compartmentalize fires when they do occur. Inspectors frequently check to see if special features, such as fire doors and separation between compartments, have been properly installed and maintained. (These points would also be mentioned in the preinspection letter.) Other fire protection features such as fire sprinklers and fire alarms are also usually inspected to make certain they are installed properly and have been routinely tested to ensure they will work when needed. But as previously mentioned, these reports can be required for review even if an inspection is not done. Notification of that fact can be provided in preinspection letters.

Local decision makers need to understand that the inspection process is more complex than just conducting inspections, identifying hazards, and correcting them. And there is more involved in the code enforcement process than their own inspection personnel and what they do.

## FIRE PROTECTION CONTRACTORS

**Fire protection contractors** are private-sector contractors who inspect, test, and maintain fire protection systems. It is very important to note that the inspection process also involves ensuring that the proper inspection, testing, and maintenance of such systems have been done according to code requirements. Not every jurisdiction monitors the reports they may require from business owners that document the inspection, testing, and maintenance of their critical fire protection systems (fire sprinklers, alarms, and fixed protection systems, including kitchen hoods and vents). Some standards that outline the process and timelines for these activities are mentioned specifically in the model codes. Others are not, but may be required by the local jurisdiction. In any case, this part of the code inspection process is critical because inspections may be done more frequently than any given jurisdiction can match with regular fire code compliance inspections.

In other words, if a jurisdiction cannot inspect a property annually, they may at least review the reports (usually called confidence reports) that document the fact that these critical systems have been inspected, tested, and maintained in accordance with the standard. And, they may follow up on any reports that show deficiencies that should be corrected immediately.

## INSPECTOR SELECTION AND TRAINING

Another major issue in administering code inspections is the need to ensure consistency. Because the codes can be complex, more experienced inspectors are often better able to identify a multitude of code violations. This can lead to complaints from occupants who state that previous inspections did not find a specific hazard, so they cannot understand why they are being "harassed" by another inspector. Consequently, inspector training is a critical need for effective management of code enforcement activities. There are a variety of training programs available for code enforcement personnel, as well as certification for both code familiarity and inspection experience. Each of the model code organizations offers some type of training and certification program for their specific code. Many jurisdictions are finding value in "certifying" their inspections through the training and testing of their personnel and close administration of their inspection process. For example (as previously mentioned), many jurisdictions provide a checklist to their inspectors to ensure a more consistent inspection. Some use preinspection letters to notify businesses what common hazards will be looked for during an inspection, so that businesses can prepare and can become partners in creating a safe environment, rather than the recipients of a forced enforcement action. But meeting certification requirements can provide an added imprimatur of reliability and credibility for the code compliance inspectors. And they most certainly ensure a higher level of proficiency than written directions that cannot take into account years of required training and experience.

Those certifications may exist at the state level, because many states offer their own inspector certification programs. They may be offered by the model code and standard promulgating organizations, such as the International Code Council and the National Fire Protection Association. But they usually have a close relationship to NFPA 1031, the professional qualification standard for fire inspectors and plans reviewers. That standard outlines the job performance requirements necessary to perform those duties and is widely recognized as a national minimum level of proficiency. Many certification tests are designed to ascertain the learning and experience that has met the level of job performance outlined in that standard.

Selecting the right kinds of people to work in a code enforcement program can be difficult, but is as important as their professional level of training and certification. Selecting firefighters may be desirable because they have experience with fire, which helps refine code development and enforcement efforts. Firefighters usually are dedicated to fire safety and can be effective advocates when enforcing the code or in persuading someone to comply. Unfortunately, many firefighters are reluctant to become "enforcers" because it means pushing people to do something they may not desire, a strong contrast to their usual role of being there to help in the event of an emergency. Consequently, some departments use nonfirefighting personnel or even volunteers to conduct inspections.

More is said about staffing patterns later in the book, but decision makers should observe applicants carefully when selecting their code enforcement personnel. Nothing can cause problems for a fire official or another local decision maker more quickly than a fire code inspector who is inept or overly attracted to the role of "enforcer." A good friend and longtime leader in the fire prevention field recently explained one of the basics of selecting the right people for the job. Wayne Powell now works for Marriott International in their safety division. Formerly, he was a program manager for fire prevention efforts at the U.S. Fire Administration. Wayne relayed a story attributed to Bill Marriott, the founder of Marriott Hotels. The company is widely recognized as having one of the highest levels of commitment to safety and customer service in the private sector. When asked how he trained people to achieve that level of proficiency, Mr. Marriott answered that he did not. He found people who already had those values, and hired them. After that factor was taken into account, training was far easier. Local decision makers would be well advised to consider the type of people they want to have enforcing their fire codes, well before hiring them and putting resources into training them.

As mentioned earlier, the inspection process is complex and requires specialized training, expertise, and attitudes for inspection personnel. In some cases, certification in the model codes requires training and passing a test, and in others, such as NFPA 1031 (the professional qualification standard for inspectors), certification requires actual time in the field.

Understanding the fire codes takes time. Add to that the need to understand the interrelationship with building and mechanical codes, and the volumes of specialty standards that NFPA produces that go hand in hand with the code, and one can easily see that expertise cannot be obtained with video training and a checklist. Local decision makers should recognize the extent of technical expertise needed for a proper inspection and take that into account when selecting and training their inspectors.

## THE APPEAL PROCESS

The final authority for fire code administration is often some type of appeals board. These boards usually are composed of architects, engineers, and design professionals from the community who have a working knowledge of the code but are not directly connected with the fire service. Their purpose is to act as the final arbitrator when the code requirements are subject to interpretation or when alternative materials and methods are suggested for equivalent levels of fire and life safety protection. One such example is when fire sprinklers are offered in lieu of other fire suppression methods. However, these appeals boards do not replace the authority of the court system when it comes to legal interpretation or legal challenges. These may be created at the state level, as in the case of the Massachusetts Commission on Fire Safety, or at the local level. They may often be combined with a building code board of appeals because of the many overlapping issues between the building and fire codes. In all cases, they present a mechanism to go beyond the local authority having jurisdiction and resolve differences in interpretation of the codes.

Managing the appeal process means establishing some kind of review board with specific authority for their task. It also means processing those appeals in a timely fashion. Board members and code management personnel should be prepared to deal with decisions in a professional manner within the scope of their authority. These boards do not generally have the authority to eliminate code requirements, but they do have broad authority to determine equivalent forms of protection. Often they are a local reality check on what would constitute an "equivalency" when business owners, operators, or contractors disagree with a fire marshal's interpretation. Managing the appeal process can be politically tricky, which means that local code officials will need to be balanced and thorough in their own administration and interpretation of the codes. Otherwise, they could be routinely overturned and gain a reputation for being too harsh, even though public safety is at stake.

Adopting codes, laws, and even policies implies that they must be administered effectively. It also implies that those involved in their administration have a responsibility to ensure that the codes they use are up to date and in concert with modern fire safety practices. To do so, decision makers at the local level must make a commitment to participate in the development of codes and standards. There is no better advocate for quality codes than the people who administer them daily, and there is no one with a better understanding of their effect in the community.

# Developing Codes and Laws: Model Codes and the Mini-Max Concept

Model code requirements must be developed somewhere. A variety of model codes currently exist that can be used as a guide for local adoption. Enforcing these codes implies some responsibility to develop and or adopt them. Not all fire departments can afford to participate in the development of model fire and life safety codes, but those who do usually participate in one of the model code processes. These currently include those produced by the National Fire Protection Association and the International Code Council. Some years ago, the Western Fire Chiefs Association (WFCA) together with the NFPA melded elements of the Uniform Fire Code into NFPA 1. It may be referred to as either NFPA 1 or the Uniform Fire Code. These national model codes are the foundation for adoption and modification at the state or local level.

Model codes are not always adopted exactly as written. Some states eliminate certain provisions of the code, leaving it up to local jurisdictions to manage issues such as fire department access or water supply for specific occupancy types on their own. In 2009 the International Residential Code (IRC) was passed with provisions that mandated the inclusion of residential fire sprinklers. However, home building interests have been fighting the adoption of those provisions within the IRC state by state, in some cases getting them excluded from the state codes. The future with regard to the IRC itself in this regard is unclear because of continued vote battles over the fire sprinkler provisions.

Fire departments with the capability to participate in codes and standards development have an important role in developing model codes based on actual field experience in firefighting and inspection activities. Participation usually involves membership in the organization, travel to their code development meetings, and voting on the language of the model code. Many organizations now allow input and proposals via the Internet, but voting is still (currently) conducted by physical presence at a meeting. This activity is not considered "essential" by many jurisdictions across the nation. Recent economic difficulties (to put it mildly) restrict out-of-town travel, and so getting field practitioners to physical locations for votes is becoming even more difficult than in the past. But if no one else participates, then only a few—those funded by business interests—could ultimately determine the makeup of the model codes.

In some cases, these model codes represent the *minimum* standards for fire and life safety, and they may be exceeded by local jurisdictions depending on specific needs or desires of a community. For example, some jurisdictions have adopted local ordinances that exceed their statewide requirements, as in the case of the fire sprinkler ordinance in Louisville, Kentucky. Normally, fire sprinkler requirements (usually determined at the state level) are intended for new construction. Louisville was able to demonstrate their inability to control fires in high-rise buildings through traditional firefighting means. As a result, they were able to pass a requirement to place fire sprinklers in all commercial high-rise buildings that was retroactive and superseded the previous building codes.

Other codes, primarily building codes, are *mini-max* in their scope. This means they represent both the minimum *and* maximum fire and life safety requirements and cannot be changed except by specific appeals or an amendment process. There are some communities that exceed model code requirements: for example, those with requirements that sprinklers be installed in all residential properties. Local decision makers should be aware of the mini-max aspect of their respective codes and the administrative procedures for local variations.

## CODE INTERPRETATION, REVISION, AND RETROACTIVITY

Throughout this chapter, the concepts of interpretation, code revision, and retroactivity have been discussed. Codes are not always black-and-white documents. Some provisions are subject to interpretation, and so may be interpreted differently by different people. An appeal process is often in place to provide some oversight for that interpretation process. Codes may be revised or even adopted differently from the model codes, depending on local circumstances. And codes are not usually retroactive (especially in building codes) because of the expense involved. The condition called *preexisting nonconforming* refers to places that were built to be in compliance with previous codes, but would not comply with current codes. The issue of retroactivity is more difficult for fire codes, because often important threats must be dealt with despite the fact that business occupancies complied with previous code editions. In those cases code is again being interpreted, and care must be taken to ensure balance. Retroactively requiring every single change in the model codes would be costly in both economic and political terms. But doing so in cases where the community clearly understands the hazards can be politically and practically palatable.

The example of changes in fire sprinkler protection for public assemblies (nightclubs) that arose out of the Station nightclub fire in Rhode Island is a perfect illustration. By law, not code adoption, many jurisdictions across the nation ordered these occupancies to retroactively install fire sprinkler systems because of the perceived threat nightclubs posed to life safety.

As fire officials participate in the code development and administration process, they will be exposed to another emerging trend affecting the process used by each of the model code organizations. As the codes are used and promoted internationally, the people who use or adopt them become motivated to participate in their development.

## GLOBALIZATION OF CODES AND STANDARDS

One of the emerging trends in fire and life safety protection is the movement of the business community toward participation in the global economy. Trade agreements between nations open up economic opportunities for international companies, who desire consistency of code applications across international boundaries. Because of this trend, there is a need for codes and standards that can be used in more than one nation. Product standards and construction practices were the first to feel this pressure, but other codes and standards are following suit. As the demand for these codes and standards increases, the number of people who participate in the adoption of the model language also increases. Consequently, code and standards development processes are now experiencing a surge of participation by other national interests with their own perspectives and histories with regard to fire and life safety. These other national interests have their own views about how codes should be written, based on their own experience and scientific analysis. Some have very different positions about how many safety features should be provided and what constitutes an acceptable level of risk. Many cannot afford the protections taken for granted by people in more industrialized nations. They also want to have their votes counted when model code language is developed.

An example was a conflict that occurred in the 1990s over standard provisions for storage of liquefied petroleum gas (LPG) found in NFPA 58. Because Mexico had for many years allowed the storage of LPG on rooftops for convenience, they did not understand the need for other measures of protection found in the document, which was used predominantly in the United States. Because they were looking to improve their own safety procedures, representatives from Mexico and the LPG industry began challenging some of the assumptions of that safety standard. A code promulgation battle ensued over what constituted truly safe practices, with many in the U.S. fire service stating that allowing changes to the storage procedures would create "rockets on rooftops." In the next code revision cycle, the standard was modified to allow storage on the roof of a structure only when the local AHJ would allow it.

For all practical purposes, that language allowed those who were used to (and comfortable with) the storage of LPG on rooftops to continue their practice without conflicting with the model standard. It also allowed AHJs who were *not* comfortable with the concept to enforce the prohibition of rooftop storage, again without being in conflict with the model standard.

That particular case is an example of how globalization of trade and communication can influence model code and standard promulgation. Consequently, those with a stake in the development of codes and standards are now being faced with other points of view from a broader international context. Decision makers will have to meet this challenge with active participation, sound science and engineering techniques, and accurate data to influence the code development process in an international arena. It is not necessarily a threat to our own code development processes to have participation from less restrictive nations, and their involvement should not be viewed as such.

## LEGAL ISSUES

Maintaining records and ensuring the completion of hazard abatement is important for legal purposes. There are numerous court cases[6] where local jurisdictions have been found liable because they failed to abate hazards that had been or should have been identified. Local authorities should also be aware that entry to inspect a property is not automatically granted and may require some kind of administrative or even criminal warrant from an appropriate court. There may be some protection for local jurisdictions from liability for failing to inspect if they do not have the resources to do so. They may also be protected from decisions that are part of their normal responsibilities. Local expert legal counsel should always be obtained to determine what level of protection from liability fire officials and decision makers may enjoy.

## Summary

Code enforcement is a principal function of a comprehensive prevention program. Fire officials and local decision makers must understand the foundation and the limits of their authority to adopt and administer fire and life safety codes. The components of the code administration process include identifying an inventory of properties to be inspected, maintaining a database and record-keeping system, selecting and training personnel, conducting inspections, and managing the appeal process.

Fire officials should participate as much as possible in the code and standards development process and recognize that the globalization of codes and standards will challenge their operations and their ability to influence code development. They should also understand the legal issues of adopting and managing a code enforcement program and obtain legal counsel that can provide guidance about the extent of their authority and their liability when enforcing fire codes.

## Case Study

### Code Enforcement and Development in the City of Wade Gardens

The City of Wade Gardens is located in the southeastern part of Florida. It serves an area of about 55 square miles, with a population of roughly 35,000 and growing. The population has nearly doubled since 1990, and Wade Gardens is projected to have 82,000 residents by 2020.

Fire protection for the city was initially provided by a volunteer fire department. In 1968, the fire department was reorganized as a branch of the city government. Today, the department has three stations, 77 uniformed personnel, four civilians (nonuniform staff), and 20 active volunteers.

The Life Safety Services Division, formerly known as the Fire Prevention Bureau, conducts activities for plan review, new construction inspections, existing occupancy inspections, prefire plans, community education, juvenile firesetter intervention, and fire investigations. The staff includes a fire marshal, three fire inspectors, a community education specialist, and a number of dedicated volunteers. The Life Safety Services Division also interacts with other city departments such as Building and Planning and Zoning on growth issues.

The fire inspectors perform various inspections, complete prefire plans, investigate fires to determine their cause and origin, and assist with community education programs. The community education specialist provides a variety of educational programs on fire safety and injury prevention and serves as the public information officer for the fire department. The volunteers assist the division in fire investigation and community education activities.

In the past year, the division conducted 4,709 inspections, which were done by both regular designated inspectors and active fire companies. That number included new construction inspections. The continued growth of the city was evident with a total of 487 sets of construction plans reviewed by the division, up 16 percent from the previous year.

Code enforcement activities had previously been governed by a Code Enforcement Board (CEB), composed of seven local citizens appointed by the City Council to operate as a "quasijudicial" body that decided whether to levy fines based on the code enforcement department's recommendations. Last year, the City Council voted to abolish the CEB and replace it with a "special master," or judge, to decide on code violations.

Part of the rationale for replacing the citizens board was the fact that it pitted "neighbor against neighbor," according to local news reports.

The function fulfilled by the CEB and the special master is that of the arbitrator for code enforcement issues. Violations are issued by the fire department, and willing compliance is sought. However, those refusing to comply or those wishing to appeal take their case before the arbitrator.

For example, in a previous year, a building owner was cited by the fire marshal for failure to have central station monitoring for the building's fire sprinkler system and fire alarm system. The building owner refused to comply with orders to correct the situation, stating that the building was not required to have central station monitoring

when it was built in 1979. A review of the code indicated that local code required central station monitoring as early as 1973.

The building owner obtained a certificate of occupancy allowing him to operate and refused to comply with the code requirements. He was ordered to appear before the city's Code Enforcement Board for a hearing, where he was found to be in violation of the code and subject to a fine of $250 per day if the violations were not corrected within 30 days. The building owner appealed the CEB decision to the county circuit court.

After 22 months of process time, the county court heard the owner's case, denied the appeal, and ordered code compliance within 15 days. The owner complied, and the case was closed this year.

# Case Study Review Questions

1. From which level of government does the authority to conduct fire code inspections presumably derive?
2. Which branch of government has the final authority to determine compliance? What ramifications does this have for the code authority within the fire department?
3. What potential difficulties would arise from having emergency response personnel conduct fire code inspections? What advantages would that type of inspection activity yield?
4. What type of research is necessary before a code enforcement case is taken before a board of appeal?
5. What advantages/disadvantages would be associated with a board versus a single person handling appeals?

6. What would have been different about the appeal had the code requirement for central system monitoring of the smoke alarms *not* been present at the time of construction in Wade Gardens?
7. What is an occupancy, and why is that knowledge important?
8. What is the difference between a code and a standard?
9. Why is record keeping such an important part of a code enforcement program?
10. What role do fire protection contractors play in the code enforcement process?

# Endnotes

1. *Fire Inspection Management Guidelines.* Quincy, MA: National Fire Protection Association and Fire Marshals Association of North America in cooperation with the U.S. Fire Administration, 1982.

2. *Conducting Fire Inspection: A Guidebook for Field Use.* Quincy, MA: National Fire Protection Association, 1982.

3. *Fire Inspection and Code Enforcement.* Stillwater, OK: Fire Protection Publications, International Fire Service Training Association.

4. *Fire, EMS and Emergency Preparedness.* Retrieved from http://www.sysplan.com/ Our%20Capabilities/Fire%20EMS% 20Emergency%20Preparedness.

5. Hall, John R., Jr., *Fire Code Inspection and Fire Prevention: What Methods Lead to Success?* Quincy, MA: National Fire Protection Association, 1979.

6. Selected court cases from Appendix A of *Fire Inspection Management Guidelines*, 1982, developed by the National Fire Protection Association and the Fire Marshals Association of North America in cooperation with the U.S. Fire Administration: *Adams v. State (of Alaska)*, 555 P. 2nd 235, (1976); *Coffey v. City of Milwaukee (Wisconsin)*, 74 Wis. 2d 526, 247 N.W. 2d 132.

# CHAPTER

# 4

# The Plan Review Process: Engineering Elements in Prevention

## KEY TERMS

Acceptance inspections *p. 76*　　Construction plan review process *p. 76*　　Performance-based design *p. 81*

## OBJECTIVES

After reading this chapter you should be able to:

- Describe the relationship between construction type and use and the features fire departments should examine during the plan review process.
- Recognize the interrelationship between the fire code and other codes.
- Describe how performance codes are different from prescriptive codes.
- Recognize the potential problems associated with streamlining plan review functions.
- Describe the appeal process, relative to the plan review for new construction.

PEARSON
## myfirekit™

For additional review and practice tests, visit **www.bradybooks.com** and click on MyBradyKit to access book-specific resources for this text!

construction plan
review process
■ activities associated
with making certain
that new construction
or remodels comply
with appropriate build-
ing and fire codes

Not all fire departments participate in the plan review process for new construction. The **construction plan review process** consists of the activities associated with ensuring that new construction or remodels comply with appropriate building and fire codes. The process includes reviewing submitted plans, then approving them or having them modified. It also involves field inspections to make sure that what is submitted and approved on plans is actually constructed at the work site.

The construction process involves a complex set of interrelated needs that requires a comprehensive view of development that matches local concerns. New construction must generally meet requirements for fire and life safety, zoning, environmental, mechanical, plumbing, and structural stability. Once constructed, buildings often must also meet the requirements of a fire code for safe use of the structure. Reviewing plans for new construction provides an opportunity to ensure compliance with safe building practices before buildings are actually constructed. The process is not merely a matter of reviewing plans and approving them. Usually, permits are not issued for actual construction until plans have been received, reviewed for code compliance, and approved. Then the process also includes what are commonly called **acceptance inspections**. Acceptance inspections often involve testing of features such as sprinklers or alarms to make sure that they are working according to design standards and code requirements.

acceptance inspections
■ physical visits done in
the field to ensure that
what is approved on
the plans is actually
built in the field

Fire departments that do not participate in this part of a comprehensive approach to fire prevention are missing an opportunity to ensure that buildings are constructed as safely as possible under modern codes. They are also missing opportunities to make sure that adequate water is available to sustain fire flow, that adequate fire department access is available, that fixed fire protection systems are in place, and that adequate egress for emergencies exists. Fire departments have an added interest in ensuring all this because the same fire protection systems that protect the building will be used by fire department personnel during fire emergencies, and the adequacy of those systems can affect the safety of the firefighters.

At best, fire departments that do not participate in the plan review process relegate that role to another agency that is checking for many more code compliance issues, including mechanical, structural, electrical, environmental, and plumbing. The likelihood of missing a problem during plan review increases with the complexity of the review.

At worst, there is no one checking construction plans, or doing field (acceptance) inspections to ensure that what appears on the plans is actually constructed. The responsibility for compliance is left solely to the business developer and the construction contractors.

Furthermore, the plan review process and acceptance inspections are important not just during new construction. In one case, the deconstruction of a building had drastic consequences, allegedly because of a lack of official oversight into the process. The Deutsche Bank Building in New York City caught fire during the deconstruction process in August 2007. Local news accounts, even years later, laid the blame on inadequate inspection of safety features.

According to an online article published by Wayne Barrett of the *Village Voice* (http://www.villagevoice.com/2009-07-22/news/bloomberg-s-biggest-scandal-the-deutsche-bank-fire-should-be-his-downfall-why-isn-t-it/), the Deutsche Bank building had so many safety problems that their consultant United Research Services (URS)

Plans are reviewed for compliance with building, fire, and mechanical codes, and especially for fire and life safety issues such as fire sprinklers, alarms, egress, fire department access, and water supply. (*Courtesy of Virginia Department of Fire Protection—Virginia Fire Marshal's Academy*)

reported to officials that the construction management firm could no longer be trusted. That notice occurred in August 18, 2007, and 15 days later a fire started by a discarded cigarette injured 115 firefighters.

Problems were also identified in the lax inspection practices by the city—which allegedly contributed to missing a 42-foot breach in the fire standpipe system.

Two firefighters were killed in that fire. This tragedy highlights how important it is for fire officials to pay close attention to both the construction plan review process and plans for deconstruction.

# The Fire Department Role in Plan Review

Some fire departments view their role in plan review as limited and relegate that responsibility to other governmental agencies in the development process. In doing so, however, departments miss an opportunity to utilize one of the strongest prevention concepts, *engineering*, to help control their fire problem and protect their firefighters. These departments rely on other agencies to make certain that structural requirements designed for firefighter safety during emergencies have been adequately reviewed and that the reviewed plans have been complied with. As important as this is, it is not the only reason for the fire service to concern themselves with plan review as one aspect of a comprehensive prevention program. Participating in the plan review process can help reduce the expense of providing fire protection after the fact and improve public safety at the same time. It can also ensure that fire department needs for firefighting purposes are met.

The process of reviewing plans requires time and expertise with the codes. Reviewers must also have some understanding of the construction process and how issues translate from a piece of paper to actual construction in the field. Typically, fire departments are concerned with development features such as hydrant spacing, water supply, fire sprinklers, and street width and other emergency-vehicle access issues. In addition, fire departments have a special concern for exiting requirements and alarm systems, which can have a direct impact on the number of people needing rescue during an emergency.

These safety factors must be examined during plan review if the fire service plans to take advantage of engineering concepts that help ensure a safe level of construction and use. The building department usually determines the use of a building and its construction type before it is reviewed for these other fire safety issues. It is important for plan reviewers *and* decision makers to understand that buildings are constructed for certain uses and that their plans must be reviewed accordingly. One construction type will not generally suit all types of occupancy use. For example, the construction requirements for a public assembly occupancy (e.g., a concert hall) usually involve more exits than a warehouse structure of the same square footage. This is because the number of people in a public assembly is far greater per square foot, and they must be able to exit quickly in the event of an emergency. Similarly, the construction type for operations that include hazardous materials is usually far more stringent because of the potential for greater harm due to the type of materials being used in the structure.

The various construction types include various engineering requirements for fire resistance and safety. More restrictive requirements might include that all structural members, including walls, columns, beams, floors, and roofs, must be made of noncombustible or limited combustible materials having a specified degree of resistance. At the other end of the spectrum, less restrictive requirements would allow exterior walls, load-bearing walls, floors, roofs, and supports to be

Field "acceptance" inspections ensure that what is constructed in the field matches what was approved on the plans. Further, they ensure that the equipment is functioning as it should. (*Starke, FL Fire Rescue*)

made wholly or partly of wood. Greater detail about construction types can be found in a variety of books, including the *IFSTA* manuals[1] dealing with fire inspection and code enforcement or building construction. They are also found in the National Fire Protection Association standards regarding building construction.

The interrelationship between occupancy type, or use of the building, and its construction requirements is rooted in past disasters that have shaped code requirements. The fire service must understand this relationship fully and be prepared to respond in an appropriate fashion, knowing that what was originally approved may not be what is found on arrival at the scene of an emergency.

The plan review and regular inspection process would be far easier if the use of a building never changed. The relationship between the building construction and its occupancy use can be affected when a new type of business locates in a building that is not designed to handle its safety needs. For example, owners of a common warehouse that stores consumable goods may decide to change their business operation to store hazardous materials, thinking that one warehouse is as good (for business) as another. Fire inspectors making regular rounds of their inspection district often discover this type of problem. Most businesses are focused on their own operational needs. Either they lack the expertise required to assess their new safety needs, or they overlook the problem entirely. They are not necessarily trying to avoid safety requirements; they may in fact be uneducated or unaware. In other cases, businesses are so concerned about the costs of construction that they will deliberately avoid the expense of bringing their structure up to code requirements when their business model changes or there is a change in ownership and operations.

If code compliance inspectors do not understand the relationship between building construction and building use, they could miss unsafe situations that should be corrected. As a result, fire department participation in the plan review

Adding high-rack storage in a warehouse is one of the most common changes that occur in business operations after a building is occupied. Without knowledge of the codes, building operators often assume that available space can be filled up and are ignorant of the more restrictive fire code requirements associated with the increased fire load.

process is extremely valuable. Communication between those who conduct regular code compliance inspections and those responsible for construction permits is crucial. At a minimum, there must also be close communication between the building and fire departments so that this interrelationship between construction and use can be effectively coordinated. The interrelationship among the various codes (mechanical, plumbing, etc.) can also be coordinated.

## FIRE DEPARTMENT CONCERNS DURING CONSTRUCTION

As previously stated, the plan review process is an opportunity to ensure that the fire department can respond effectively in an emergency. Fire department vehicles need to be able to position themselves effectively. The water supply must be sufficient to handle a fire (which means calculating fire water flow requirements), and an adequate number of fire hydrants must be placed in the right locations for fire-fighting purposes.

Fire sprinklers are currently part of the fire code; but in future (as in the past), they may be regulated in the building code. Sprinklers are usually installed during the new construction process, unless significant structural modifications are being made to an existing building and it must be brought into compliance with newer codes. In any event, fire departments should have an active interest in sprinkler installation. Fire sprinklers have proven themselves to be effective about 96 percent of the time (according to the National Fire Protection Association) at extinguishing a fire or at least controlling it until the fire department can respond. Consequently, fire sprinklers are now required in many types of buildings.

Different types of buildings require different sprinkler systems. Some systems are designed for special applications, such as dry sprinkler systems for areas where water pipes might freeze. Other sprinkler systems are specially designed for use in homes. Sometimes fire sprinklers must be retrofitted into older buildings being remodeled. All of these factors related to fire sprinkler types should be examined during the plan review process. It is often valuable to have dedicated professionals, who understand the complexity of engineering the construction and hydraulics of a sprinkler system, available to make sure the sprinklers are the correct design and are being installed properly.

Exiting requirements are also important to the fire service. It has proven impossible in the past to rescue large numbers of people from a fire when they are trapped inside. Having exits that are wide enough is not the only factor to consider for safety. The materials used to construct the exits are equally important, because these escape paths must continue to work even when there is a fire elsewhere in the building. This is why more fire-resistant construction, self-closing doors, and emergency lighting are often required in exits. Exit spacing (distance between exits) and marking are also critical issues. The Station nightclub fire in Rhode Island in 2003 highlighted this particular aspect of exiting, which is still being debated. Although other emergency exits existed for this particular club, most of the people killed in the fire were stacked up at the front door, trying to exit where they had entered the establishment. This phenomenon may have implications for future code requirements for building exiting and for educating patrons.

To be sure, there were other problems with that particular fire. The illegal use of indoor pyrotechnics for a rock-and-roll show inside started the fire. The rapid spread of the fire was fueled by illegal foamlike panels placed on the ceiling and walls to improve acoustics. But the lack of fire sprinkler protection and the problems already identified with exiting behaviors contributed to the heavy losses in that fire. Reports and analysis by the National Institute of Standards and Technology have led to recommendations that front exits be expanded to handle two thirds of the occupant load by themselves, despite requirements for other exiting.

Issues that may be found (or missed) during a regular code compliance inspection often are referred to those responsible for new construction permits, plan review, and acceptance inspections associated with that part of fire or building code compliance. Consequently, there is yet another necessary relationship between those responsible for plan review and the code or standard development process. The link between building construction, use, and maintenance and code or standard promulgation has a nexus during the construction process. It is during new construction (or significant remodeling) that improvements in fire protection features are the most affordable. Retrofitting may be expensive, so the plan review process is where many modern fire protection features are dealt with. But understanding the link between construction and a building's use is critical to code or standard development.

The issues that arise during the construction plan review process are complex. For these reasons, many fire departments are requiring a higher level of professional training and certification for their employees involved in plan review. In some cases, they are recruiting and hiring fire protection engineers to make sure that expertise is available to properly evaluate plans, and especially to manage **performance-based design** elements.

However the plan review process is accomplished, the fire service must understand, and other local decision makers should recognize, that the fire service has an active interest in the building construction process, and therefore in reviewing plans for fire and life safety issues. Adequately staffing plan review efforts can be a significant challenge during tough economic times. But staffing levels of a fire department can be affected by the need to provide for large rescues when exit requirements have not been adequately addressed. And failure of key fire protection systems can put firefighters at risk as well as the public they serve. Thus, paying attention to these issues and ensuring adequate staffing are critical.

## FIRE DEPARTMENT CONCERNS DURING DEMOLITION

Fire departments should also be concerned with fire safety during demolition. As previously mentioned, in the 2007 New York City fire in the Deutsche Bank building, which killed two firefighters, safety features not in place during demolition, such as a disabled standpipe system, contributed to the fire. As with new construction and the hazards that may be present before fire protection features such as fire sprinklers are in place, fire departments need to be concerned for the same reasons with buildings that are being torn down.

## SPECIALIZED ISSUES

Plan reviewers must also be able to handle other problems associated with certain types of construction or use. For example, when urban growth expands into wildland areas surrounding a community, several unique firefighting problems arise. Urban sprawl places demands on an urbanized fire department that may not be prepared to fight large-scale wildland fires. Under these circumstances a review

With more fire departments using fire protection engineers for their plan reviews, the international Society of Fire Protection Engineers—a nonprofit that exists to support and promote the field of fire protection engineering—offers a valuable resource. (*Courtesy of the Society of Fire Protection Engineers*)

**performance-based design**
■ design based on goals for fire and life safety rather than relying on the prescribed requirements found in the model codes

for vegetation control, roofing material requirements, and additional water supply is also necessary to identify other ways to control fire losses at the interface between urban and wildland areas. Generally speaking, greater water supplies are required, and wider roads for fire department access are needed to allow people to escape while fire agencies are responding to the scene. On-street parking presents a special hazard in these situations when people escaping in automobiles meet responding fire crews on a street too narrow to handle the traffic load during such an emergency.

Other special situations require specialized reviews. Hospitals require different safety features than schools because of the type of people using the structure: most school children can easily move, but patients confined to a hospital bed cannot. Businesses that use large amounts of hazardous materials will have specific fire protection requirements that may be unfamiliar to smaller jurisdictions not used to handling them. Development of a large petroleum or liquefied petroleum gas storage facility presents unique problems that are complicated to review; the review process (including acceptance inspections) can be extremely time consuming.

There are other situations that require close attention by the fire service. However, whether or not they are actually reviewed by fire service personnel, these construction issues are always of concern to a fire department when considering the entire fire protection and loss control package that makes up the fire protection capabilities of any community. It is therefore valuable to have fire department participation in this process because of the interrelationship between a building's construction and its safe use. Also, because the nature of these requirements changes for different uses, special expertise is often required during the plan review process.

It is critical that the fire service be involved in the plan review process. Monitoring important fire protection features for many specialized buildings can save lives, including those of firefighters—especially where there is a possibility of large fires and explosions, such as at grain storage and loading facilities. (*Courtesy of Vancouver Fire Department, WA*)

When the complexity of code requirements is beyond the capabilities of those managing the plan review process, additional help may be necessary. Some jurisdictions find it valuable to contract with (or hire) fire protection engineers to assist them with the plan review process. Some may relegate the assessment of plans and their approval to a third party who can ensure that code requirements have been met. Often that assurance requires an engineer's stamp. Many private firms in the development industry are hiring their own engineers to provide technical reviews of construction plans or to assist in design. If those doing the plan review and approval are not capable of assessing a technical review with an equal measure of expertise, they may be ignorant of critical flaws and miss important safety items during the plan review process. Issues of coordinated or conflicting codes also highlight the need for specialists and an adequate level of technical expertise during the plan review process.

## INTERRELATIONSHIPS BETWEEN CODES

Portions of the construction process are generally regulated by specific codes. Each of the model code development groups has separate provisions for fire and life safety, structural stability, electrical, plumbing, and mechanical features. They are usually split into specific codes and standards, which are interrelated to some degree to ensure cohesive and consistent application. How one code is managed can affect another. For example, the mechanical code would primarily govern heating and air conditioning. Much of the attention paid to these functions will be concerned with the cost of the energy they require. But the movement of air in a building can also affect smoke distribution during a fire. So there may be fire safety requirements (such as smoke dampers that prohibit air movement during a fire) in either a fire or a mechanical code—or both.

Much of the electrical code exists to prevent fires. The plumbing code may deal with special water quality issues caused by the installation of fire sprinklers, such as backflow devices that keep stagnant water in a fire sprinkler system from fouling domestic water supplies. Also, some plumbing codes include specialized information of concern to the fire service, such as codes relating to medical gas plumbing, which often constitutes a unique fire hazard and a danger to responding firefighters.

A special level of expertise is generally required for each separate code. But an understanding of how they work together is critical.

There are many codes that may further complicate the construction process. Different codes for planning, zoning, transportation, environmental concerns, sewers, storm drainage, and water supply combine to make a complex situation for development and produce challenges for fire protection. Environmental and planning codes may conflict with fire protection issues where natural vegetation is concerned. Generally, environmental proponents want natural vegetation in place to help with soil stabilization and water quality and to create a natural environment for animal life. But that interest may conflict with the need to create defensible spaces around structures to help minimize dangers during a wildfire.

The building code generally establishes the prescriptive requirements in building construction for safety purposes. However, in certain occupancies and in different portions of the country, these fire and life safety concerns are found in the NFPA *Life Safety Code*.[2] Both the NFPA and the ICC have building and fire codes. But the

Not all fire issues are addressed in the fire code. The mechanical code regulates heating, ventilation, and air conditioning features of the construction process. But air movement can be critical for smoke management in a fire scenario. (*Courtesy of McKinney Fire Department, TX*)

International Residential Code produced by the ICC governs construction requirements for one- and two-family dwellings in most of the nation. So the residential construction code could supersede the requirements of a regular building code.

A modern building code will include provisions to ensure that the building is structurally sound. However, various environmental factors can change these requirements. For example, earthquake stability is generally more of a concern in the western United States. The southeastern states have more specialized requirements for high winds. And considering snow loads on structures is obviously much more important in the northern states than it is in the South.

Planning codes usually delineate density and zoning requirements. These issues can have a direct impact on the need for fire department emergency response capabilities. For example, the planning department will usually be the agency that decides whether housing can be built in a wildland area. They will also determine housing densities and the location of business or manufacturing operations, which require specific fire department response capabilities. Housing density can be critical, because placing houses closer together increases the hazard of fire spread.

Transportation codes cover road construction requirements and may sometimes conflict with fire department goals for quick emergency response. The requirements for adequate fire department access are often addressed in these codes, and access in urban or wildland interfaces is only one of the potential problems that may arise. For example, the desire to calm traffic by reducing road widths or erecting speed bumps to reduce traffic accidents can create delays in emergency response. The need for traffic calming and the need for emergency vehicle access are both legitimate and are usually supported by different community sectors that are concerned about one type of problem more than another. It is somewhat ironic that a fire department may be interested in traffic calming to reduce auto pedestrian emergencies, while being concerned that the traffic calming devices will slow response should an emergency occur.

Environmental codes may exist separately or may be encompassed in broader codes that deal with sewage disposal, water supply, and water drainage. Fire departments have obvious concerns about water supply for firefighting purposes. But here, again, fire protection issues may be affected by other interests. Vegetation control may directly or indirectly compete with water quality and environmental concerns. Consequently, some fire departments participate in environmental impact reviews to determine how fire hazards can be mitigated while coordinating with other environmental issues. Creating natural collection basins for storm-water runoff can provide an attractive hazard that could affect a community's drowning rate. And these basins can create large natural areas that inhibit fire department access.

Further complicating the plan review process, and highlighting the interrelationship between the various codes and standards, are the standards themselves. A critical aspect of the plan review process is determining *how* systems will be installed. As previously mentioned, the National Fire Protection Association publishes a wide variety of standards that are frequently referenced when installation and safety practices are needed. For example, *NFPA 13*[3] deals with installation and performance standards for fire sprinkler systems. *NFPA 72*[4] does the same for fire alarm systems. These standards are most often referred to by plan examiners to hold contractors accountable, ensuring proper installation. In addition, the model code organizations for building, fire, and other construction-related concerns also have standards of their own. Other organizations, such as Underwriters Laboratories and Factory Mutual, also develop more specific standards for installation and product performance. The Consumer Product Safety Commission (CPSC) also monitors performance of some safety materials, such as fire sprinklers and alarms, and occasionally intervenes to ensure public safety. One such intervention by the CPSC created a nationwide recall of fire sprinkler products that were identified as faulty after being used for years in many businesses.

The interrelationship between codes and standards is complex, but can generally be described in simple terms. Usually codes tell us what needs to be done, and standards tell us how to do it. But having the expertise to understand how all the various codes and standards work together, how they are related to one another, and how the interests behind them may conflict is a critical part of the plan review process.

Within the plan review process, there is another issue that requires some attention. It is the effort to aid the development process by allowing specific designs for construction and development that meet performance goals, rather than prescriptive requirements.

# Performance-Based Codes

Many fire codes allow the use of alternative materials and methods to achieve an end result that is equivalent to prescriptive code requirements. For example, fire sprinklers often are accepted as an alternate method of providing adequate water for suppression when a developer wishes to reduce water flow from hydrants or fire department access requirements. Performance-based codes take that concept and apply it in general.

Performance-based construction design allows for alternative ways of reaching those goals. Currently, model fire and building codes usually contain specific requirements for safety features. Fire sprinklers, smoke alarms, fire separation between floors of a building, and structural elements to provide earthquake stability are a few examples. Performance codes establish goals to deal with those safety issues that allow alternative solutions and a design process that is specific to the hazards of the building and its location, rather than following specified requirements.

Efforts to achieve performance objectives rather than prescriptive code requirements provide even more opportunities for creativity in crafting local solutions to construction problems. However, the movement toward performance-based codes produces additional challenges for fire officials and local decision makers. When codes are performance based, it is advantageous to have fire department personnel involved in the plan review process so that engineering solutions can be balanced against emergency response capabilities. For example, a fire department that does not have enough personnel to adequately handle a high-rise fire may want increased levels of protection to mitigate the potential problems of a fire. But in some (rare) cases, developers have actually proposed constructing spaces in some buildings that would be protected by building compartmentalization rather than fire sprinklers—and making use of emergency response capabilities as part of their protection plan.

Fire Marshal Joseph M. Fleming of the Boston Fire Department has faced problems that illustrate what others may experience when dealing with performance codes. A developer once proposed a performance-code design for a high-rise building that initially did not include fire sprinklers. Fleming has (in numerous speaking engagements) highlighted the problems that local jurisdictions face with performance-based designs, which often involve computer modeling, because they lack the expertise to evaluate them. The computer models rely on assumptions that may be challenged by those with expertise in this area. Most computer models are still in their infancy and may have flaws that another design professional could identify. For this reason, local officials may want to require a third-party review by experts in the field to accurately determine whether performance objectives will actually produce the desired safety results. In addition, other local officials have hired fire protection engineers as permanent staff to help evaluate performance-based designs.

Sometimes the prescriptive code does not even address a particular situation, and performance of fire protection features must be evaluated without any standard requirements to which they can be compared. For example, creating a paint spray operation large enough to handle a modern airliner may involve creating fire protection features that are unique. Provisions for paint spray booths or areas found in the code may be inadequate to address the size of operations needed for such an activity.

Under any of these scenarios, the complexity of performance-based design, which involves accepting alternative methods or materials rather than adhering to a prescriptive code, will require specialized knowledge and expertise.

To the extent that fire departments already have experience in exchanging some protection features for others, performance-code issues have presented an opportunity to look at local solutions from more than one point of view. Still, performance codes represent a significant challenge for fire departments. They require a level of expertise (e.g., engineering and computer modeling) that most fire departments do not have. Fire officials and local decision makers will have to

Fire laboratory tests are used to gather data that can be used by computer software programs to model fire behavior. This type of information becomes very important when considering alternative solutions—especially in the performance code arena. (*Courtesy of Hughes Associates, Inc.*)

ensure that they can provide adequate scientific and engineering analysis of performance codes in the plan review process, or require that a third party provide that service as part of the plan review process.

# Streamlining the Development Plan Review Process

The pressure to streamline development procedures for local jurisdictions often leads decision makers to conclude that a "one size fits all" plan review process will be of great benefit to the economic vitality of a community. This is not an unreasonable point of view. Maximizing the use of staff and resources for plan review does make sense. And in smaller jurisdictions, where all the buildings being constructed are relatively small and plans do not involve complex reviews, some level of expertise in several different codes may be possible.

Although consistency and efficiency are legitimate goals, caution should be a guiding factor when designing a municipal or county system to review plans for construction and development. Given pressure to streamline the review process and make it more efficient, it is possible to forget that codes and ordinances for construction are usually in place for legitimate safety purposes. Attempting to group them together in a quick review can mean that vital safety issues are missed in the plan review process. In addition, competing interests within the community that helped develop the various codes can create conflicts that may require individual advocacy as they are balanced for the collective good. The conflicting points of view between environmental or transportation codes and fire codes could be set aside by an individual not familiar with them, or the process could be torn between those competing interests.

Developers may want increased density within planning codes to allow for more development. Local decision makers may desire the same thing to prevent urban sprawl, a problem that has a natural citizen constituency within many communities. However, local property owners may want more space and resist further density for construction, particularly for housing. Environmentalists desire more trees and vegetation to protect water and air quality, while fire departments are attempting to reduce fuel load for wildfires that can devastate a community, even in urban areas.

In addition, as previously addressed, the various codes can be complex and can involve conflicting interests and content. It would be impractical, if not impossible, for one person to know all the details of building, plumbing, mechanical, electrical, fire, planning, environmental and transportation codes. A lack of advocacy for fire safety issues in this complicated environment could lead to disasters such as fires with large loss of life: disasters that capture the attention of the community and drive them to ask how such a situation could have been prevented.

On December 3, 1999, six firefighters lost their lives fighting a fire in a cold-storage warehouse. On June 18, 2007, nine firefighters died battling a fire in a warehouse in Charleston, South Carolina. In such cases, it is not just firefighting tactics that come into question. There will also be heavy public scrutiny about code compliance, the plan review process, and who was responsible for making sure the buildings were constructed properly.

When a disaster occurs, people (and the media especially) want to know how it could have been prevented. And when codes or standards existed to provide protection, but were not followed, the public will want to know who is responsible.

Competing interests in the construction process can create problems for the fire service. Saving construction costs by limiting street widths can create fire access issues. (*Courtesy of the McKinney Fire Department, TX*)

# The Appeal Process

The appeal process during plan review is usually the same as it is for regular code enforcement. In fact, many jurisdictions have established an appeals board that serves both regular fire code inspections and the plan review process concurrently. They may also have one board that reviews appeals challenging both fire codes and building codes. Chapter 3 provides more information on this subject. But local decision makers should be aware that the plan review process itself is much more fertile ground for appeals, because of the complexity of the construction process and the many codes involved. Contractors and developers scrutinize the process intensely, constantly monitoring the time it takes to obtain service. They are generally well aware of code provisions, or they have hired professionals with a certain (sometimes high) understanding of the codes to help them design their projects. It is not uncommon to have a code interpretation challenged by an architect or a contracting professional who works out of a specific NFPA standard every day. These people can be extremely knowledgeable about code and standard provisions.

Any code official trying to bully or buffalo a developer or contractor into submission with provisions that are not clearly spelled out in the codes or standards is taking a serious risk. Developers and contractors are often influential within a community, and have close ties to political leaders, because they are constantly involved in the growth that is so crucial to the survival of many local tax bases.

When performance codes are used, or when the costs of prescriptive requirements are high, these developers or contractors are highly motivated to search for less expensive alternatives. And they will often apply whatever political pressure they can to encourage liberal interpretations of the code.

It is critical that local code officials and decision makers set up and manage an appeal process that both provides public accountability and protects public safety. Decision makers should select people to serve on an appeals board who have some technical expertise and are familiar enough with at least some of the codes to understand the concepts behind them. Appeals board members should be able to help evaluate complex engineering solutions for fire protection problems, and they should understand the legal consequences when codes are ignored.

## Summary

Plan review is a critical part of a comprehensive prevention program. It provides a fire department with the opportunity to engineer safety designs into its community and prevent fires rather than fight them after they occur. Plan review is the process of reviewing construction plans for code compliance, issuing permits to begin construction, performing acceptance inspections to make sure that what is built matches approved plans, and finally allowing occupancy of a newly constructed (or significantly renovated) building.

Plan review requires specialized training and expertise, and it must be done in a coordinated fashion with other codes. The complexity of the codes (plumbing, electrical, mechanical, and building, etc.) makes coordination challenging but critical. Streamlining the development plan review process can be difficult because of those complex issues.

Being involved in plan review represents significant challenges to fire officials, particularly when alternative materials and methods are proposed as some part of a performance-based safety objective.

But missing the opportunity to be involved could mean that safety features are missed, are installed incorrectly, or are not working as designed. If a fire occurs under those circumstances, the public will be asking questions about how those problems could have been resolved with adequate attention to the plan review process.

## Case Study

### Prevention Efforts in Uthe, Arizona

Uthe, Arizona, is a (fictional) city serving about 1.3 million people spread out over 473 square miles. The fire department has 45 fire stations and more than 1,250 sworn personnel, with about 70 assigned to the Prevention Division. The Prevention Division is responsible for directing all the prevention activities for the fire department, including fire investigations, plan review, public education, and code enforcement.

Within the scope of prevention activities, the Uthe Fire Prevention Division has responsibility for fire and life safety plan reviews and works in conjunction with their building department. The Prevention Division conducts regular fire code compliance inspections in existing occupancies with a staff designated specifically for that purpose. Fire crews do not conduct fire code compliance inspections, but volunteers from fire crews do provide fire safety inspections for state-licensed facilities such as day care centers and group homes. Volunteers also respond to service requests and perform "educational inspections" where the goal is to gain willing compliance.

The Plan Review Section is responsible for working with architects, engineers, contractors, business owners and managers, homeowners, and the City of Uthe Development Services Department to ensure that buildings constructed within the city are safe and that the materials associated with these buildings are handled, stored, and used in a safe manner. The fire code permitting process can be divided into four modes of operation: consulting with customers, plan review and inspection, code development, and code enforcement. The consulting program involves working with customers to solve their problems. The plan review and inspection program is the formal permitting process. The code development program involves updating the building and fire codes to address issues raised by new or changing processes and procedures and correcting problems with the existing codes. The code enforcement program ensures that required permits are obtained, construction site inspections are conducted, complaints are investigated, and fire-lane parking and the city's smoke ordinances are enforced.

The Fire Safety Advisory Board is a citizen's group consisting of 14 members who are residents of the City of Uthe and represent various industries, trades, and professions. The board may make recommendations on matters pertaining to the Uthe Fire Code or ordinance provisions and amendments. The board also hears appeals of the decisions of the fire chief as part of the administrative hearing process.

The Plan Review Section is also responsible for adopting and maintaining the Uthe Fire Code, currently an amended version of the International Fire Code produced by the International Code Council. Additional duties of this section include certification of pyrotechnic operators, blasters, and transporters; updating the Fire Prevention Division's operation manual; and coordinating training for the Fire Prevention Specialists.

Recently, a local developer wanted an interpretation on a high-rise building that would allow the elimination of some of the fire protection features. At issue was their own code interpretation, suggesting that if the fire marshal would designate a particular location for fire department access, the building would not be tall enough to be considered a high rise. Under that scenario, the building would include fire sprinklers, but pressurized stairwells for egress (those that help to keep smoke out of the egress path) would be eliminated. In addition, establishing a command center for fire department operations (used for communication throughout the building) would likewise not be required.

The fire and building codes require a measurement from the lowest point on grade to the top of the building to determine if a building is designated as a high-rise under the code. The developer has arranged a meeting with the city manager and the fire code official to present an alternative location for ladder operations, and a proposal to eliminate the code requirements.

# Case Study Review Questions

1. What are the advantages to linking the plan review process within the Fire Prevention Section with code development and adoption for the City?
2. What advantages are there to having the plan review efforts (for new construction) closely linked with fire code compliance activities in existing buildings?
3. What problems might arise from the use of volunteers for educating owners about fire code compliance in lieu of a regular code compliance inspections?
4. After examining the case study, describe a course of action for the fire marshal to respond to the request of the developer to eliminate requirements associated with a high rise.
5. What is the importance of consulting with customers during the plan review process? Describe the kinds of help that customers (building developers, designers, and construction professionals) might need.
6. What is the difference between a performance and a prescriptive code?
7. What are some potential problems with streamlining plan review processes?
8. What are acceptance inspections, and what role do they play in the construction process?
9. Why is a concern for building safety *during* construction or deconstruction so important?
10. What problems frequently arise regarding high-rack storage situations?

## Endnotes

1. International Fire Service Training Association, *Fire Inspection and Code Enforcement*. Stillwater, OK: Fire Protection Publications; International Fire Service Training Association, *Building Construction Related to Fire Service*. Stillwater, OK: Fire Protection Publications.

2. *NFPA 101: Life Safety Code*. Quincy, MA: National Fire Protection Association (NFPA).
3. *NFPA 13: Standard for the Installation of Sprinkler Systems*. Quincy, MA: National Fire Protection Association.
4. *NFPA 72: National Fire Alarm Code*. Quincy, MA: National Fire Protection Association.

# 5

# Fire, Arson, and Explosion Investigations

## OBJECTIVES

After reading this chapter you should be able to:
- Describe the differences between fire cause investigations for unintentional and those for purposely set fires.
- Describe the importance of data collection and the preservation of evidence.
- Describe the rights, responsibilities, and legal limits of an investigator.
- Link cause-and-origin investigation to the community's fire prevention program.

A critical part of any comprehensive prevention effort is the information on which it is based. The data used to plan successful prevention efforts comes from the active investigation of factors that led up to the incident. Notably, these "incidents" are not called *accidents,* because they are usually caused by foreseeable events. The active investigation of these factors will provide the important "what happened" information that is the foundation of preventing them from occurring again.

Most fire departments investigate fires to identify the area of origin and determine the probable cause, though the level of sophistication with which they do so varies widely. Sometimes the cause is readily evident, and responding fire officers can successfully conclude how a fire originated. A fire caused by a lightning strike, with accompanying accounts from credible witnesses, requires no further investigation. But for most fires, the cause is not so clear. The basic purpose of investigating a fire, therefore, is to help determine its cause so that similar fires can be prevented in the future. A complete fire investigation may also identify other factors that led to the start or the expansion of a fire. For example, apartment doors that had been left open—encouraging the rapid spread of a fire beyond the area designed to keep it in check—would be an important finding even though they were not necessarily part of the cause of the fire.

**arson**
■ the crime committed when an individual has deliberately set a fire

Fires are also investigated to determine whether the crime of **arson** has been committed. If a crime is suspected, the nature of the investigation changes, and law enforcement officials become partners in the investigation procedure. In a criminal investigation, the scope of the investigation process usually increases dramatically. The police and public prosecutors may take an active interest in the investigation, and interviewing witnesses may become more complex.

Accelerant detection dogs are often used as a first step to see if a fire has been deliberately set. The Bureau of Alcohol, Tobacco and Firearms has a model program to support this service for some local fire departments. (*Courtesy of Kingsport Fire Department, TN*)

This chapter is devoted to five important aspects of the investigation process: fire **cause determination**; the preservation and documentation of evidence; the crime of arson and its unique requirements for fire investigation; the use of data from the investigation process; and legal issues associated with the investigation process.

# Fire Cause Determination

Even when a crime is not suspected, there are elements of an investigation that may involve outside interests. For example, most insurance companies routinely hire their own investigators to determine the cause of a fire. Insurance companies investigate fires to protect themselves from liability claims or to recover costs from responsible third parties through subrogation suits.

Subrogation suits are legal actions designed to recover costs through court determination of the responsible party (or parties). For example, the manufacturer of an electrical appliance that caused a fire may be sued by the insurance carrier to recover expenses for the damages of the actual fire and the expenses of the investigation. In such cases, private investigators hired by insurance companies may or may not agree with local authorities about the cause. And when a very expensive fire loss occurs, special interests with a stake in the financial loss can make the investigation complex and subject it to scrutiny from many outside sources. These circumstances demand highly proficient fire investigators who can present solid scientific evidence to support their conclusions about cause. This legal aspect of fire investigations is driving many fire departments to increase their levels of proficiency, in order to produce neutral findings that are not biased by business or financial interests.

Some activities are common to any investigation, whether a fire occurs through unintentional events or is purposely set. According to *A Pocket Guide to Arson and Fire Investigation*,[1] the investigation of a fire scene actually begins before the fire is extinguished. Responding fire crews may make observations that will later help investigators determine how the fire began. For example, the color of the smoke and flame can reveal a great deal about the nature of the fire, to what extent it had burned before suppression began, and even what type of materials were burned. These observations should be shared with a trained investigator. According to the *Fire and Arson Scene Evidence: A Guide For Public Safety Personnel*,[2] first responders may also observe other factors, such as the presence, location, and condition of witnesses, or vehicles leaving the scene.

Once the responders have arrived and suppressed the fire, most investigators recommend that overhaul procedures (conducted to make certain the fire has not spread) should be delayed until after a preliminary investigation that includes a physical examination of the scene and the initial interviews of witnesses, including firefighting personnel, to help recreate the fire scenario.

The suppression personnel who arrive first are in the best position to see where the fire was first concentrated. They may also receive information from witnesses who leave the scene before investigators arrive. Suppression personnel should pay particular attention to signs of arson, such as blocked or locked windows, multiple fires, and sabotaged fire protection features. They should also note any suspicious people or comments and relay these to investigators. There is, of course, a great deal more to this part of an investigation, and an

Examining burn patterns can help pinpoint areas of origin, leading to a cause determination. Expertise and caution are necessary—as this information may be misinterpreted or misused in fire cause or arson cases. (*Courtesy of Montgomery County Fire & Rescue, MD*)

excellent resource is the training manual *Fire Investigator,*[3] published by the International Fire Service Training Association (IFSTA). Another excellent text on the subject is *Kirk's Fire Investigation,*[4] by John De Haan and published by Prentice Hall, which is frequently used as part of the certification process for fire investigators.

Beyond what normal company personnel may observe, the science of fire investigation entails a great deal of training and expertise. That expertise begins with an understanding of the scientific principles of fire behavior. It requires skill in interpreting fire patterns, because understanding how materials react in a fire can tell an investigator a great deal about what happened. Because it is frequently possible that the fire was deliberately set, nearly every fire scene should initially be considered as a crime scene (in terms of evidence preservation).

At the same time, investigators should never assume too much. Making assumptions tends to focus investigators' attention on data that supports their theory and leads them to ignore or reject facts that do not fit the theory. *NFPA 921* recommends the use of the scientific method, which incorporates collecting data, analyzing it, then developing a hypothesis and testing it.[5] If a probable hypothesis is not developed, the cause of the fire should remain undetermined, according to NFPA 921.

However, there is a conflict at the local level regarding this approach. The effectiveness of fire investigation activities is often measured by whether a cause is found. Investigators are naturally concerned that a high percentage of "undetermined" causes could lead observers to believe that the investigators are not competent. In simple terms, what would be the rationale for providing fire investigation services if the cause for 70 percent of incidents was listed as "undetermined"? Some jurisdictions have tried to achieve a balance by listing the causes as "probable." There is some controversy over this approach within the fire prevention community, so local decision makers will have to decide on their own what kind of balance to achieve. It is an important issue, particularly because listing a cause as fully determined could lead to a subrogation suit or even an arson prosecution if the determination resulted from inadequate science or investigation techniques.

Gathering data is therefore critical and includes a physical examination of the fire scene. Many aspects of the fire scene provide an investigator with valuable information. For example, examination of burn patterns can lead investigators to the area of origin, even in a badly damaged structure. Burn patterns of ordinary wood can help indicate whether accelerants were used to start a fire. Multiple points of origin would also indicate that a fire was deliberately set. Burn patterns are based on the length of time materials have been exposed to fire and the directions in which fire spreads. These burn patterns are observed and interpreted by comparing areas where fire destroyed or altered material and where it did not. Patterns can be observed on structural elements, wall surfaces, and furnishings.

This study of the pattern geometry can direct investigators to the area of origin and, consequently, help in efforts to determine the probable cause. The origin of the fire can be determined using interpretation of fire patterns, careful collection of witness information, and analysis of the data collected.

For example, narrowing the likely area of origin may lead to a kitchen scene, and even more specifically, a particular appliance. However, providing evidence that a fire occurred in a particular appliance may require some preliminary understanding of how the appliance operates and ultimately a scientific analysis conducted by a private laboratory. Sometimes the destruction is so complete that there is insufficient physical evidence to establish a cause. Conversely, there are times when a single item of physical evidence or dependable eyewitness information can be the basis of a conclusive determination.

There are, of course, a wide variety of causes for fire. Often it is caused by some electrical malfunction. Carelessly discarded cigarettes are also a principal cause of fires and fire deaths. Many fires also occur in the kitchen because of poor cooking practices, such as leaving food unattended while it is cooking. The cause is always some form of heat source coming in contact with fuel that will burn. In addition, human actions, deliberate or unintentional, are usually a significant factor in a fire's cause or spread.

Determining the cause of a fire and the relationships among all the factors that contribute to it can be a daunting task that requires specialized expertise. Because it is the mission of a fire investigation program to *accurately* determine a fire's cause, those performing the investigation must be well trained to recognize fire burn patterns and apply the science of investigative techniques to the physical characteristics of the many materials inside a fire scene. Training is not enough. Adequate fire investigation efforts also require the equipment necessary to identify causes and aid in prosecution of crimes. For example, many fire investigations require laboratory reviews to corroborate an investigation's conclusion. Many departments also use accelerant-detection dogs. These are specially trained dogs that "alert" (signal) in the presence of hydrocarbon-based products. Properly trained dogs are able to detect extremely low levels of flammable liquids. Samples can then be taken for laboratory testing and confirmation. Ultimately, this information can aid in the case for arson.

One of the seminal publications on the topic of fire and explosion investigations is the *NFPA 921* published by the National Fire Protection Association. It provides a great deal of information about the process of an investigation, including the behavior of fire and burn patterns. It also outlines the procedures for the collection and **preservation of evidence**, which are discussed later in this chapter.

preservation of
evidence
■ collection and pro-
tection of accurate
notes and physical
evidence

Interviewing witnesses to the fire is also a critical part of fire investigation, for which investigators must be trained. Witness interviews can help recreate the fire scenario, which is particularly important when arson is suspected. However, similar techniques are used whether arson is suspected or not. The best training for interview techniques will probably be found through local, state, or regional police associations, because police deal with witness interviews more frequently than do fire personnel. However, many fire investigation associations (such as the International Association of Arson Investigators) also conduct high-quality training, including programs for successful interview techniques and courtroom testimony.

When arson is suspected, sorting out the cause of a fire can be misleading and extremely difficult. Motive, opportunity, physical evidence, and criminal interview procedures all come into play. Because the investigation process goes beyond cause determination, partnerships with police, prosecutors, and fire department personnel are highly recommended. If arson is suspected, a criminal investigation and all the rules associated with it are in effect. Some fire departments actually train and certify their investigators as police officers. However the partnership is managed, particular attention must be given to the area of cause determination for arson or suspected arson cases, because of their complexity. More is said about this aspect of fire investigation later in this chapter.

# Evidence Preservation

The IFSTA *Fire Cause Determination, NFPA 921,* and *Kirk's Fire Investigation* all deal more extensively with collection and preservation of evidence than does this text. According to the IFSTA *Manual,* evidence may be direct, circumstantial, or physical. Direct evidence is composed of facts, which can be reported firsthand. Circumstantial evidence is an inference formed from more direct evidence. Seeing someone with matches immediately before a fire started would be circumstantial evidence—inferring a relationship between the person and the fire without firsthand observation. Physical evidence may include burn pattern geometry, chemicals, containers of flammable liquids, or other tools left behind by an arsonist.

Preservation of these types of evidence takes on different forms. *Preservation* refers to the collection and protection of accurate notes and physical evidence—and ensuring that the chain of evidence is unbroken in case it is needed in court. Documenting an unbroken chain of evidence ensures that it has not been tampered with.

Notes taken by the investigator are a crucial part of the investigation process. Careful collection of direct and circumstantial evidence is handled through detailed notes, which include witness information, times, dates, sketches, maps, pictures, and other pertinent information. The collection of physical evidence involves careful handling and storage so as not to contaminate the evidence with outside influences (such as the investigator's fingerprints). All evidence must be carefully preserved and a clear chain of evidence established. This chain is necessary because the purpose of all types of evidence is to serve as the official record of events, whether in court or in insurance proceedings.

# Arson Investigations

Motivations for deliberately setting a fire may be very different depending on the person who is setting it. Generally, firesetters' motivations fall into the following categories:

- Fraud (direct and indirect) for a variety of reasons, among them to collect on insurance
- Pyromania (serial firesetter)
- Spite or revenge
- Crime concealment
- Civil unrest or terrorism
- Juvenile-set fires
- Vanity, as in volunteer firefighters or security guards who wish to gain credit for discovery or early suppression attempts

## JUVENILE FIRESETTERS

Most of the identified motivations for deliberately setting fires are associated with adult behavior. But fires are also often set by children who do not have any actual intent to produce harm or profit.

In some cases, children are merely curious, setting a small fire that then gets beyond their control. In other cases, they are acting as a result of some emotional disturbance, but still do not intend serious harm. In either case, a screening interview is usually conducted to determine the child's motivation and to make sure he or she is referred for proper treatment. Curiosity firesetters are usually given an educational intervention and placed under close supervision to prevent further events. Those who are acting from an emotional disturbance require more treatment and are usually referred to mental health professionals who specialize in such cases. Screening tools for interviewing juvenile firesetters are available from the U.S. Fire Administration, among other sources. The screening tool was developed by mental health professionals to help differentiate between normal curiosity firesetters, and those with more emotional problems that lead to firesetting. These screening tools are really guided interviews with accompanying score sheets. They should be used only by personnel who have been trained to use them effectively. Screening tools are actually useless if the proper mental health services are not available for troubled children once they have been identified. Untreated or severe emotional disturbances in young firesetters can lead to more severe firesetting behaviors. It is very rare to run across a person who fits the psychological profile of being a "pyromaniac," or someone fixed on fire with deep emotional and psychological problems. But untreated firesetters can escalate their deliberate fire sets, causing increasing fire damage and even death.

Then there are others who deliberately set fires. Those motivated by profit or revenge fall into a different category of firesetting behavior. In some cases, even young children may have such motivations and may be charged with the crime of arson and dealt with by the court system. Obviously, adults usually know better, and when a fire is deliberately set, a criminal investigation of the act should occur.

Proving arson can be very difficult because of a lack of physical evidence. During arson investigations, the relationship among investigators, the local police, and prosecutors is critical. A report done for the U.S. Fire Administration by TriData Corporation[6] concluded that because the activities of cause determination, arson investigation, and the development of a criminal or civil case occur in sequence, close coordination between fire and police agencies is particularly important. In some cases, fire departments train and certify their fire investigators as police officers to more effectively control the investigation sequence in criminal cases. In any case, the chain of evidence must proceed from the initial investigation through the entire prosecution system.

Beyond the actual investigation and its technique, the results of a fire's investigation can contribute to the database of information available to help guide prevention efforts.

# The Use of Data from Fire Investigations

The use of data in the fire investigation effort involves compiling and analyzing data. This data can be statistical or anecdotal, and both types present valuable information that helps educators design proactive prevention strategies.

**statistical data**
■ the historical bits of data that provide information about patterns of fire cause

**Statistical data** refers to the accumulation of fire incident and cause information. Tracking a history of fire incidents and assembling the cause data over time will yield information about the principal causes. The U.S. Fire Administration and most U.S. states use the National Fire Incident Reporting System (NFIRS) to collect fire incident information. This helps create an accumulation of data that can be analyzed for root causes, and also used to compare fire causes in different parts of the nation. This type of data is very important for targeting prevention efforts where they are needed most.

**anecdotal data**
■ personal examples, informal witness reports, news articles, etc., that are more limited, but often more revealing because they can represent one particular case in human terms that captures attention of decision makers

But where data does not exist, or is too limited to provide valuable information, **anecdotal data** can be compelling and aid prevention efforts. An anecdotal example would be more of an individual observation. Providing an anecdotal example on even a single incident can be revealing. Sometimes, going beyond statistical data to collect anecdotal information can provide insight into *why* fires may be set—what set of contributing factors yielded the results. Often, the fire investigators are the best source of anecdotal information. For example, they might be among the first to identify patterns of fires caused by candles, and the motivation behind them. In one jurisdiction, an increase in religious ceremonies accompanied by burning incense, and sometimes drug use, resulted in an increase in fires caused by unattended candles. Alternatively, in-depth research into an individual incident may reveal a problem with faulty equipment that could more appropriately be handled by changing the manufacturing process. That depth of information may not show up on a statistical report, but investigators were able to identify root causes through an understanding of individual—or anecdotal—examples.

Identifying cause information is important in planning effective prevention measures. Gathering data is critical to provide us with that information—but without analysis of the data we collect, our planning efforts can be misdirected. For example, attacking an increase in candle-set fires without understanding the root motivation behind their use would likely miss the mark. Therefore, analysis of the data is part of the fire investigation process.

Whether conducted by investigators or analysts, this function of a fire investigation effort is critical in helping design preventive solutions. A thorough investigation and analysis process might reveal that papers left too near a portable heater, doors left open to let air feed the fire, and the flammability of wall coverings *all* contributed to the fire cause and its rapid advance. Specific examples such as these can bring the complex nature of investigations to a simple level that laypersons can readily understand.

In addition to anecdotal evidence, long-range evaluation of loss data should be part of the goal of an accurate fire investigation program so that prevention strategies benefit from the historical perspective a good investigation process provides. Consequently, the record-keeping system for fire investigations is a critical long-term tool of comprehensive prevention efforts.

The National Fire Protection Association and the U.S. Fire Administration both produce reports from an accumulation of data and investigation of significant individual fires. These reports provide fire services with positive examples of how to assemble their own statistical and anecdotal data to guide their prevention efforts. Often the data and reports lag "real time" because it takes sometimes years to accumulate the data. Following are some examples from the report *Fire Loss in the United States 2007,* by Michael J. Karter, Jr., of the Fire Analysis and Research Division of the National Fire Protection Association.

### Overview of 2007 U.S. Fire Experience

#### Number of Fires

- 1,557,500 fires were attended by public fire departments, a decrease of 5.2% from the year before.
- 530,500 fires occurred in structures, an increase of 1.2%.
- 414,000 fires or 78% of all structure fires occurred in residential properties.
- 258,000 fires occurred in vehicles, a decrease of 7.2% from the year before.
- 769,000 fires occurred in outside properties, a decrease of 8.5%.
- What do these fire frequencies above mean? Every 20 seconds, a fire department responds to a fire somewhere in the nation. A fire occurs in a structure at the rate of one every 59 seconds, and in particular a residential fire occurs every 76 seconds. Fires occur in vehicles at the rate of 1 every 122 seconds, and there is a fire in an outside property every 41 seconds.

#### Civilian Fire Deaths

- 3,430 civilian fire deaths occurred in 2007, an increase of 5.7%.
- About 84% of all fire deaths occurred in the home.

- 2,865 civilian fire deaths occurred in the home, an increase of 11.0%.
- 365 civilians died in highway vehicle fires.
- 105 civilians died in nonresidential structure fires.
- Nationwide, there was a civilian fire death every 153 minutes.

### Civilian Fire Injuries

- 17,675 civilian fire injuries occurred in 2007, a decrease of 7.8%. This estimate for civilian injuries is on the low side, due to under reporting of civilian injuries to the fire service.
- 14,000 of all civilian injuries occurred in residential properties, while 1,350 occurred in nonresidential structure fires.
- Nationwide, there was a civilian fire injury every 30 minutes.

### Property Damage

- An estimated $14,639,000,000 in property damage occurred because of fire in 2007, a highly significant increase of 29.5% from last year. This total figure includes the California Fire Storm 2007 with an estimated property damage of $1,800,000,000. Excluding the California Fire Storm, total property loss still increased a significant 13.5%.
- $10,638,000,000 of property damage occurred in structure fires, excluding structures associated with the California Fire Storm.
- $7,546,000,000 of property loss occurred in residential properties.

### Intentionally Set Fires

- An estimated 32,500 intentionally set structure fires occurred in 2007, an increase of 4.8%.
- Intentionally set fires in structures resulted in 295 civilian deaths, a decrease of 3.3%.
- Intentionally set structure fires also resulted in $733,000,000 in property loss, a decrease of 2.9%.
- 20,500 intentionally set vehicle fires occurred, no change from a year ago, and caused $145,000,000 in property damage, an increase of 8.2% from a year ago.

The data presented in a report such as this one is critical, of course. However, the way in which the data is presented is often just as important, because it helps readers to understand what the data means. If a picture is worth a thousand words, then a graph of the data can provide a snapshot view of the meaning of the data and help decision makers formulate their prevention strategies.

# Legal Issues in Investigation

Local decision makers should be aware that conducting investigations also produces its own set of potential legal liabilities. Legal issues in fire investigation activities usually involve standard police rights and responsibilities about conducting criminal investigations. However, the right of entry could also become an issue for fire investigators.

Generally, courts have upheld the right of a fire department to enter a structure to fight a fire. Some departments are mandated by state or local laws to conduct fire investigations—and if these investigations are begun at the time of the fire, some departments have asserted that the right of entry has been granted to continue the investigation after the fire has been extinguished. However, vacating the premises and then returning would almost certainly be viewed as relinquishing the property back to the owner—and therefore legal warrants to reenter the property would be required to finish the investigation. That is why most fire departments do not relinquish control of the "fire scene" until the physical investigation has been completed.

In the 1984 U.S. Supreme Court case *Michigan v. Clifford*, the Court attempted to define a continuation of an investigation and clarify the circumstances that would require a warrant. In that case, some evidence was ruled inadmissible because control of the property had been relinquished to the property owner. This case and others have generally found that if investigators maintain possession, they can actively investigate. After control has been relinquished, an administrative or criminal warrant would then be necessary to reenter the property. For those reasons, the investigation procedures done at the time of the incident should be thorough and well documented both in writing and with photographs. The physical collection of evidence is also critical at this time.

Some states (Washington state, for example) do not have administrative warrants. Only criminal warrants are available for search of private residences, so the type of information needed is more extensive, and a violation of law must be fairly evident in order to obtain a criminal warrant for reentry. For these reasons, local decision makers must thoroughly research the investigation laws under which they operate, and proper legal counsel should be obtained and consulted.

Local decision makers should ensure proper investigator training and establish clear-cut policies and procedures to protect themselves from legal challenges by ensuring a measure of quality review and providing ongoing training. Certifications for investigators are available and demonstrate a higher level of professionalism—which can often be helpful in court cases about deliberate fires, or subrogation suits where faulty equipment may be a cause. Decision makers should also obtain local legal expertise to review their practices to ensure compliance with appropriate local, state, or federal requirements.

Recently, more jurisdictions have found that they have a legal obligation to provide for investigator safety. Although investigators are not (as part of their function) inside a building when it is on fire, they are subjected to the hazardous atmospheres that are the residual effect of a fire. They are also subject to some of the same biohazards that firefighters may face—for example, at a drug-lab fire scene there may be dirty syringes, which can puncture skin and expose fire investigators to disease. Local decision makers must understand these hazards, and provide adequate protection for their personnel or face legal challenges related to employee safety.

# Fire Inspections Law Bulletin

*How fire marshals and fire code inspectors use the law to protect their communities.*

| Vol. 8, No. 1 | FireChiefLaw.com | January 2010 |

## In This Issue

### Attempt to Block Code Inspection – Criminal Charges

The business owner *refused* to allow the inspector and an accompanying police officers entry to conduct the code inspection. After entry is forced, the owner became abusive and was finally arrested and charged with obstructing legal process and assault. ................................................................................................................... Page 2

### Exposed Wiring – Apartment Stove – Fire Causes Serious Injuries

After filing suit, apartment management and owners obtained an order for summary judgment. On appeal, Plaintiff looked to her son's eyewitness testimony to establish the fire's *origin and cause*. Is such testimony sufficient?........... Page 3

### Fire Hazard – Parents Charged With Criminal Mistreatment

The house had no operable smoke alarms, and all means of egress were blocked. The electrical circuits were overloaded and flammable debris was piled high throughout the house. The parents are arrested and charged. ........................... Page 5

### Public Nuisance – Fire Damaged Home – Repairs Not Completed

After obtaining a building permit to repair the damage, the city issues a *stop work order* because the homeowner did not complete the repairs within the limit set by the city. What is the homeowner's remedy, if any?........................ Page 6

### Fire Caused by Discarded Cigarette? – Expert Witness Not Required

Expert witnesses are required to determine the *origin and cause* of a fire only when the assumed cause of a fire is beyond the common experience of jurors. In this instance, the factors were not complex enough to require an expert witness. ................................................................................................................................... Page 4 sidebar

### Domestic Dispute – Intentionally Set Fire

There was sufficient evidence that the suspect argued with her boyfriend and used an accelerant to *intentionally* set a fire. ................................................................................................................................... Page 6 sidebar

### Annual Index

Index of all cases reported during 2009. Arranged by topic. ........................................................................... Page 8

---

## In The Next Issue

### Owner Accumulates $400,000 in Fines – Constitutional Violation?

The city ordinance provides no appeal process for second and subsequent fines for the same code violation. Here, the owner claims his *due process rights* were violated because most of his fines were from second and later violations.

---

EDM Publishers Inc. (www.edmpublishers.com) • PO Box 2423 • Duxbury, MA 02331 • T: 800-859-6402 • info@edmpublishers.com

One valuable commercial service is a newsletter for Legal Concepts in Fire Investigation. It is a case study tool to help local managers stay abreast of legal issues around the field of fire investigation and fire inspections. (*Courtesy of EDM Publishers*)

# Summary

Fires are investigated to determine the cause. However, the process used for investigations changes once arson is suspected. Under those circumstances, the criminal aspect of the investigation puts the rights and responsibilities of investigators into another legal realm. The partnerships necessary to successfully identify cause and to make a legal case for prosecution involve police and prosecutors at a minimum. Under any circumstances, the collection and preservation of evidence is critical—and maintaining an unbroken chain of evidence is necessary to demonstrate that evidence has not been tampered with.

Juvenile firesetters are an important element of investigation procedures, because of the nature of their motivation. If they are just curious, then an education intervention may be all that is required to get them to stop setting fires. However, if underlying emotional problems are motivating firesetting behaviors, then another course of treatment will likely be required—and in extreme cases, prosecution may be involved. In any case, some type of psychological screening tool is necessary to determine if emotional problems were a factor in the firesetting—and adequate counseling resources must be made available to prevent repeat offenses.

The data obtained from fire investigations is critical to planning for fire prevention efforts. Either statistical (historical) data or anecdotal (individual case) data may be combined to create a true picture of the primary causes of fires. This causal information should then be used to help target prevention programs toward the people, their behaviors, and any other contributing factors that may prevent fires from occurring. In some cases, faulty equipment can be identified through proper investigations, and a prevention effort could be mounted to fix the equipment or have it banned for use.

The legal ramifications for fire investigations are evident in the procedures used for gaining entry into a site to conduct investigations. Generally, if property is kept in the possession of a fire department during a fire, courts have upheld the department's right to conduct an investigation. Relinquishing control over the fire site could mean that investigators must obtain warrants for reentry, complicating matters greatly. In any case, local decision makers should consult local legal expertise to make sure that federal, state, or local ordinances regarding the rights of entry, interrogation or interview, or any other aspect of the investigation process are followed.

Local decision makers also have a legal obligation to provide for fire investigator safety and should be thoroughly aware of the rules surrounding protective equipment, including self-contained breathing apparatus, and the exposures that investigators encounter while conducting fire investigations.

# Case Study

A fire erupted in the Earlstone apartments in the early morning hours of a summer day. The fire department arrived to find one apartment fully involved with fire, threatening apartments on either side. Luckily, the apartment was on the upper floor of a two-story apartment complex, so no apartment was directly overhead. Multiple alarms were called to assemble the necessary equipment and personnel to fight the fire. The fire department for that community did not maintain a 24-hour investigator, and so the on-call investigator was notified by pager that she would be needed on scene.

When the on-call investigator arrived, she noticed that the fire crews had controlled the fire and were beginning to overhaul the building to make sure that the fire was out completely. There was extensive fire spread into the attic area, and the on-duty battalion chief was concerned about the amount of resources needed to hold the fire in check so that the investigator could begin an interior examination of the scene.

The observations of the first in fire crews helped pinpoint the apartment where the fire originated, but entry to the apartment would mean donning full protective gear, including self-contained breathing apparatus. Concurrently, several people in the apartment complex said that they had first-hand information for the investigator, including one who said he was suspicious of another neighbor. The investigator was trying to decide whether to enter the structure first, or to interview people at the scene. She knew she needed help, but there was a budget restriction in place for overtime, and she was uncertain whether to call out another investigator, which would require more overtime.

She consulted with supervisors, and after some time another investigator arrived on-scene to help with the investigation. Ultimately, the apartment residents were found and interviewed, and their information was compared to what was physically observable at the fire scene. The likely cause was determined to be combustibles that had been placed too close to a high-intensity lamp. There had been no smokers in the apartment, and the neighbor who was suspicious about another's activity had no information to connect that person with the time or location of the fire. Subsequent interviews with others corroborated that there was no connection between that individual and the fire scene—and in fact revealed that there was a personal conflict between the individual mentioned and the suspicious neighbor.

Damage estimates were listed at $350,000, but the insurance company actually paid out more than $500,000 in claims related to the construction damage. None of the residents of the apartment complex had insurance on their belongings, and the family that lived in the apartment where the fire started ended up losing everything.

## Case Study Review Questions

1. What kind of problems does it present for the fire department to maintain control of the scene while waiting for an on-call investigator?
2. What kind of challenges does it present to hold up fire suppression activities while the investigator conducts an interior examination of the fire scene?
3. How could issues about having a second on-call investigator been handled differently?
4. What legal ramifications could result from not having enough investigators to do both interior examination and interviews of persons of interest at the same time?
5. What is the importance of linking cause determination to ongoing fire prevention efforts?
6. What is the importance of preserving evidence?
7. What is the difference between a fire cause investigation and an arson case?
8. What are subrogation suits, and why are they important?
9. What is the importance of NFPA 921 for fire investigations?
10. Name three motivations for setting fires.

# Endnotes

1. *A Pocket Guide to Arson and Fire Investigation,* 3rd ed. Factory Mutual, 1990.
2. National Institute of Justice. *Fire and Arson Scene Evidence: A Guide For Public Safety Personnel* (research report). June 2000. Available at www.ncjrs.gov/txtfiles1/nij/181584.txt.
3. Fire Investigation, 2nd ed. International Fire Service Training Association, 2010.
4. De Haan, John D. *Kirk's Fire Investigation,* 6th ed. Upper Saddle River, NJ: Prentice Hall, 2006.
5. National Fire Protection Association. *NFPA 921.* Available at www.nfpa.org.
6. See TriData report, *A View of Management in Fire Investigation Units,* vols. 1 (1990) and 2 (1992). Federal Emergency Management Agency and the U.S. Fire Administration.

# 6
# Research in Fire Prevention

## OBJECTIVES

After reading this chapter you should be able to:
- Describe pure and applied research.
- Describe the applications of fire prevention research being conducted.
- Correlate the relationship between human behavior research and fire safety and prevention.
- Identify organizations conducting fire prevention research.
- Describe the value of research for fire prevention programs.

PEARSON
**myfirekit**

For additional review and practice tests, visit **www.bradybooks.com** and click on MyBradyKit to access book-specific resources for this text!

# Defining Research

What is research? A standing joke on this topic describes taking information from one source as plagiarism, but taking it from several as research. There is an element of truth in that joke—but it is important to understand what we are dealing with when we talk about research and its value for fire prevention programs.

Dictionary.com defines research as the "diligent and systematic inquiry or investigation into a subject in order to discover or revise facts, (or) theories." This definition is important to the field of fire prevention because too many decisions are made on the basis of individual judgment, or cumulative political will, rather than evidence or research. In recent years, leaders in the fire prevention field have been advocating more **evidence-based decision making** and improving the level of science and reason being applied in prevention efforts. *Evidence-based decision making* refers to the process of gathering information and analyzing it before making decisions. For example, in fire prevention program design, doing a search for available research information on smoke alarm placement, designing public education messages for particular audiences, and reviewing local investigation reports on smoke alarm performance are all examples of gathering information relevant to program design in advance. Decisions are based on evidence, rather than intuition or personal bias.

Practicing evidence-based decision making means targeting programs where evidence indicates a relationship between program efforts, a particular audience, and a reasonable chance at a successful outcome. It also means advocating for more research and improving the sophistication of that research for fire prevention efforts.

A simple example of what *not* to do would be developing a fire safety program for senior audiences that used hand puppets and cartoons. Available research on age-appropriate educational materials would discourage us from this approach, because it is most likely the wrong way to reach seniors. Hand puppets would normally be more appropriate for children. This might not strictly apply if we were dealing with an adult or senior audience that was functionally illiterate. In that case, simple pictograms would be appropriate. However in any case, it is the relationship between research and program design—or evidence-based decision making—that demonstrates how research is applied in the context of developing fire prevention programs.

So, were do we find this research? Evidence—or research—from our own fire investigation programs and data collection is a good beginning. This information can help us make informed decisions about targeting our prevention efforts where they would produce the most good. Developing child-friendly prevention programs, when our experience investigating fires and the data we collect from them indicates that we should be targeting adults, is another example of failing to use evidence-based decision making. As mentioned in previous chapters, fire investigation and the collection and analysis of our own local data on incidents is a form of research that helps guide our prevention efforts appropriately.

However, in this chapter, we discuss a more formal application of the term *research*. It implies a level of scientific inquiry we may not be able to obtain or afford on our own. In this context, research provides a scientific and systematic way of obtaining information that will be valuable for our prevention efforts and can survive the scrutiny of peers when they question the validity of our reasoning.

evidence-based decision making
■ the process of gathering information and analyzing it before making decisions about fire prevention program design

The U.S. Fire Administration makes research available on a wide variety of topics of value to fire prevention professionals every year. (*Courtesy of the U.S. Fire Administration*)

## Loss Measures in Residential Structures
### (3-year average (1996–98) from NFIRS data)

| Measure | All Residential Fires | Residential Heating Fires |
|---|---|---|
| Dollar Loss/Fire | $11,271 | $9,179 |
| Injuries/1,000 Fires | 48.0 | 28.9 |
| Fatalities/1,000 Fires | 7.7 | 5.7 |

## 15-Year National Trend in
## Heating Fires in Residential Structures
### (NFIRS data and NFPA Annual Surveys, 1984–98)

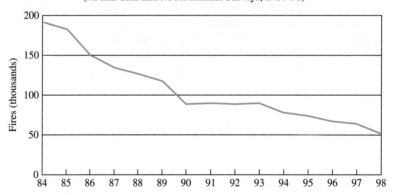

## Leading Equipment Involved in Heating Fires
### (residential structures, adjusted %, 3-year average (1996–98) from NFIRS data)

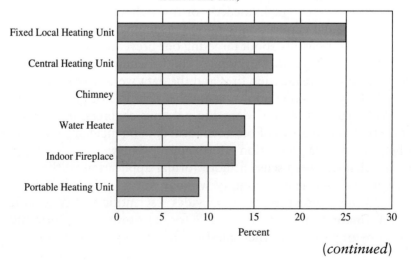

*(continued)*

## Heating Fires

(residential structures, adjusted %, 3-year average (1996–98) from NFIRS data)

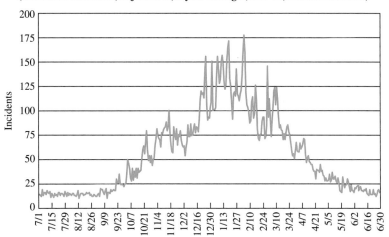

## Material Ignited in Heating Fires

(residential structures, adjusted %, 3-year average (1996–98) from NFIRS data)

| Type | Percent | Form | Percent |
|---|---|---|---|
| Resin/Tar | 28 | Rubbish/Trash | 28 |
| Sawn Wood | 18 | Structural Framing | 13 |
| Combustible Liquids | 6 | Fuel | 10 |
| Fabric | 8 | Electrical Wire | 8 |

## Leading Ignition Factors in Residential Heating Fires

(residential structures, adjusted %, 3-year average (1996–98) from NFIRS data)

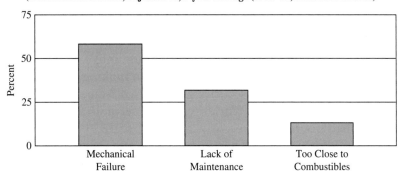

**Pure research** involves fact-finding and verification of theories for the sake of doing so. Discovering the heat release rate of burning wood is an element of pure research. Understanding that fire itself is actually rapid oxidation of a material, where heat, a combustible material, and oxygen come together to produce fire, is an outcome of pure research. **Applied research** is used to solve a particular problem. Pure research indicates how smoke can be detected through technology. Applied research can help guide our efforts to create a smoke alarm that is affordable, is durable, and provides people with enough time to escape a fire should one occur in their home.

**pure research**
- involves fact finding and verification of theories for its own sake

**applied research**
- research used to solve a particular problem

Both pure and applied research are necessary if the field of fire prevention is to move forward, and if we are to become more scientific in our methodologies to prevent fires or mitigate their effects when they do occur.

There are other applications of fire research that we need to understand as well.

# Human Behavior and Fire Prevention Research

There are two general applications of pure or applied research that are important to our fire prevention efforts. One field of study deals with human behavior. This application of research covers a wide range of topics that may affect decisions about prevention programs.

There have been many research studies on the performance of humans in fire scenarios. An excellent in-depth article on the subject, by Norman E. Grone, Ph.D., was published in the 2002 edition of *Fire Protection Engineering Magazine* (issue 16). In the article, Dr. Grone outlines a number of behavioral studies that deal with human performance in fire situations, including work by Dr. Guyene Proulx and Dr. John L. Bryan that provides insight into how people react in fires.

Conclusions in the studies referenced and from individual case studies have revealed that people may not actually panic in a fire situation—at least not at first.[1] Evidence from an earlier study by Dr. John L. Bryan, Professor Emeritus in the Department of Fire Protection Engineering, University of Maryland, indicates that people will actually help one another in a fire situation. One such study detailed how people would help one another, even strangers, to find a safe place of refuge during a fire. Dr. Proulx's research has helped dispel the widely held view that panic always ensues from a fire. She differentiates between stress-induced decision making and outright panic. She documented a body of research that disputes the notion of panic in a fire—which helps to shape building codes, particularly performance codes where evacuation is concerned.

On the other hand, studies of individual cases such as the E2 nightclub disaster in Chicago on February 17, 2003, bring into further question the notion of panic as opposed to stress-induced decision making. In that event, in an allegedly overcrowded nightclub, security guards used Mace to break up a fight. Because the 9/11 attacks on the World Trade Center and anthrax scares in the nation's capital had occurred so recently, the smell scared patrons, who may have feared another terrorist attack. They rushed to escape. There were some claims of locked exits. But, as in other cases, most people tried for the exit they knew best—the one they had used to come in. Twenty-one people were killed by trampling and suffocation in this event. Whether or not their decisions were based on stress, or rose to the level of panic, might be a factor in determining why people would ignore other exits in favor of the one they used to enter an establishment.

That issue of *Fire Protection Engineering* also mentions studies on the speech intelligibility of smoke detector systems. Today there are questions about the ability of smoke alarms to wake small children from sleep—causing major concerns about their decibel levels and appropriate teaching activities associated with children and parents. Some stipulate that if small children cannot be awakened, then interconnected alarms designed to give parents enough time to rescue them would be critical. There are also questions about the types of smoke alarms that provide the best protection. A great body of research is available, but controversy still exists as to whether ionization smoke alarms or photoelectric smoke alarms provide the best level of protection overall. Currently, the National Institute of Standards and Technology has issued no changes in their recommendations that both types of alarms are acceptable for consumer use. However, applied research on the topic exists and continues. This research is more technologically driven and is mentioned later in this chapter. But the relationship between human behaviors and smoke alarm technology is important to note.

A quick Internet search reveals that there are other research projects (market research studies) that can help determine the appropriate way to reach certain audiences. For example, The Home Study Safety Council develops educational literature for functionally illiterate audiences.

There are also a number of simple research projects conducted as part of the Executive Fire Officer program at the National Fire Academy that deal with human behavior and fire prevention topics. Examples of these research projects can be accessed from the U.S. Fire Administration's Web site. The point is that human behavior can affect many aspects of fire prevention programs. In addition, a great body of research is available to help us make informed (or evidence-based) decisions in this area. We should not be making choices based on our own bias, but rather on the evidence we can identify to help us make more informed decisions.

# Technology Research and Fire Prevention

Fire prevention technology research spans several topics relevant to the field of fire prevention. Human behavior is one aspect, but studies of fire behavior and fire-control tactics can also be applied to research into the best ways to prevent fires and to control them when they do occur.

Much of the research on fire protection technology involves fire sprinkler and smoke alarm products. These two foundations of fire protection technology have been in existence for many years.

## SMOKE ALARMS

According to ideafinder.com, the first commercially produced smoke alarm (battery powered) was designed in 1969 by BRK Electronics. During the 1970s, smoke alarms were promoted and sold heavily by BRK and others, eventually saturating the marketplace with affordable smoke alarms that were cost effective for individual homes. The technological research that made these products available and affordable also led to different designs for smoke alarms. Two such technologies were ionization-type smoke alarms and photoelectric alarms. Put simply, ionization-type smoke alarms worked on the principle that electrically charged ions in a sensing chamber would be altered by smoke and flame by-products—a detectable change that could then alert consumers about a fire in time for them to escape. Photoelectric technology relied on a sensing chamber where the smoke and fire by-products would break a visible beam of light, thus setting off the alarm and again providing early enough detection for people to escape.

Fire lab testing can be for pure research—like flame spread characteristics—or applied research to see what products may slow flame spread. (*Courtesy of Hughes Associates, Inc.*)

Over the years since smoke alarms were first introduced, fire loss statistics in the United States have dropped drastically. In the 1970s, about 12,000 people were dying each year in fires. Now that figure has dropped dramatically, to about 3,500 people dying from fire each year. Most experts agree that affordable smoke alarms have been a significant factor in producing that reduction in U.S. fire deaths. But some controversy exists.

There is an ongoing argument among fire prevention professionals about which type of technology is more effective. Some argue that ionization alarms are not as effective, because they do not alert as quickly in a slow, smoldering fire, whereas photoelectric alarms tend to alert more quickly in such fires. These proponents argue that most fatal fires are slow starting—hence the push for photoelectric technology over ionization. Dual-chamber smoke alarms are available commercially that meet the Underwriters Laboratories (UL) standards for performance. There has been considerable research on this topic conducted by (among others) the National Institute of Standards and Technology. The controversy at this point continues over which technology is best, based in part on disagreements in the scientific community about the findings of the research to date.

## FIRE SPRINKLERS AND FIRE EXTINGUISHERS

In the mid-1800s, perforated pipe systems were used in mills in New England as a means of fire protection. They were not automatic fire sprinkler systems, which were first experimented with around 1860. The first automatic fire sprinkler system was patented by Philip W. Pratt of Abington, Massachusetts, in 1872. Since then, those basic fire sprinkler systems have evolved into highly specialized forms of fire protection.

There are specialized automatic fire sprinklers that deal with certain commodities, such as highly stacked combustibles or racks of tires. There are others that use chemicals instead of water to protect certain products, such as computer systems. Most recently, some companies have been experimenting with water mist systems as a way to control fires without producing a great deal of water damage. The research associated with fire protection technology has helped the industry maintain a high percentage of successful operations. Current industry standards indicate successful fire sprinkler operations occur about 96% of the time—and most failures occur because of either human error or an inadequate sprinkler system for the type of material protected.

A company called Factory Mutual has historically conducted some research in this area that adds to this knowledge base and to the claims of human error in fire sprinkler system failures. More recently, their technical research also points out the environmental impact of fire sprinkler systems—and the supposition that fire sprinklers may actually be better for the environment than traditional firefighting.[2]

As an example, when commodities in storage change, a fire sprinkler system that was correctly designed for its former use might be inadequate for the new commodity. Hence, a sprinkler failure might occur that is not the result of a product failure, but rather of a change in the conditions under which the system had to function.

At around the same time that fire sprinklers were first being developed, portable fire extinguishers were also being invented. Alanson Crane is credited with a patent on the first fire extinguisher in 1863. Just as other fire protection technology evolved, so too did fire extinguishers. Fire service practitioners are familiar with class A (wood), B (flammable liquid), and C (energized electrical) fires. Different fire extinguishers have been designed to handle each of these fire types. Recent innovations have allowed the development of class K fire extinguishers, which are designed to control fires in deep fryers that use much more flammable oils (vegetable oil) than previous versions.

The Marriott Corporation has done some extensive testing of water mist fire sprinklers—which may prove to be most effective at limiting water damage in fires. (*Courtesy of Wayne Powell of Marriott International*)

## WHERE FIRE TECHNOLOGY RESEARCH IS HEADED

The mainstays of fire protection technology have been around for many years, and research has allowed the evolution of design improvements that match changing conditions, or even improve performance. But what of future technology? Where are the research projects and inventions that will deal with some of the continuing fire problems we face today?

There are a variety of products just beginning to enter the marketplace that may have an impact on the fire problem. Battery-operated candles are readily available and are obviously safer that wax-and-wick versions. Because candles are a principal cause of unintentional fires, safer candle design could help reduce fire incident rates. Many versions exist that have a **testing laboratory listing** and can be found with a simple Internet search.

**testing laboratory listing**
■ a stamp that indicates a particular product has met the performance requirements and safety standards for that product

Another technology beginning to enter the scene is a kitchen device that can either control a stovetop fire should one start, or prevent one from occurring in the first place. There are many such devices available, including chemical fire extinguishing systems. The National Institute of Standards and Technology is conducting tests to determine their effectiveness, but it is not yet clear how effective such systems might be.

Other nations have developed technology that shuts cooking stoves off with a timer, or a motion sensor that can tell when cooking food has been left unattended. To my knowledge, only one company, Pioneering Technology, currently produces a device that regulates the cooking temperature (for electric ranges) with a computer chip. The device, called the Safe-T-element® Cooking System, has demonstrated an ability to prevent fires from occurring while reducing energy consumption.

The Japanese are reported to have some experience with thermocouples that will shut off a gas stove by checking the temperature of the cooking pan on the stovetop. Research on this particular topic was difficult to find, and a task group working under the auspices of Vision 20/20 is attempting to identify needed research parameters and to publish the results.

This photo of a kitchen fires is a familiar example to those in the fire service. There are several promising technologies that would prevent these on an electric stovetop—notably those produced by pioneering technology for their ability to regulate heat and prevent fires from occurring in the first place. Public education is still a viable prevention option, and other technologies may exist—but NIST is examining several technologies for short term prevention strategies and research.

It is difficult for producers of new technology to obtain testing laboratory listings for their products, because of the expense involved. Moreover, with stovetop systems it is even more difficult, because the product's effectiveness with each individual brand of cookstove would be evaluated separately, which increases expenses exponentially. In other words, finding technology that will work for electric coil-top, gas, ceramic-top, *and* induction technology stovetops may ultimately prove to be impossible—and because of the increased scope of research for the different types of stovetops, the research will likely too expensive for individual manufacturers.

Those involved in the fire prevention field should understand that research into new prevention technology is an important field of study that can help make our communities more fire safe. There are limits to funding for this type of research,

A testing laboratory listing usually is demonstrated by a stamp or label that indicates a particular product has met performance requirements and safety standards for that product. National testing laboratories such as Underwriters Laboratories, Intertek, or Factory Mutual will develop safety standards for some products, and then perform rigorous tests to ensure that products will operate safely and as designed. *The "Standard in Safety" logo is a trademark of Underwriters Laboratories Inc and is used with permission. All rights reserved.*

and it is particularly difficult (because of the expense involved) to develop tests and obtain testing laboratory listings for new products. However, technological advances will continue—and there will always be a need for more research in this area.

# The Organizations Conducting Fire Research

There are many organizations conducting fire research. Some are conducting pure research, and others applied research. Some are focusing on human behavior, and others are focusing on technology. Some are highly scientific in their approach, and others are using a more rudimentary level of research methodology because of funding limitations. Staying abreast of the type of research being conducted will keep fire prevention practitioners on the forefront of new ideas and information that can improve our efforts. Some of the principal organizations that are actively involved in fire protection research include the following:

- The National Institute of Standards and Technology (NIST) (www.nist.gov)
    - The building and fire research laboratory at NIST is considered to be the predominant fire research facility in the United States.
- The U.S. Forest Service (www.fs.fed.us)
    - The Forest Service conducts research projects that focus on wildfire protection. Topics include the development of defensible space (clearing of brush) around homes that might be in the path of wildfires, and building-construction features (including external sprinkler systems) that might protect structures from wildfires.
- The U.S. Fire Administration (www.usfa.dhs.gov)
    - The U.S. Fire Administration maintains a library of resources at the Learning Center of the National Fire Academy (NFA). There are also a number of research projects conducted as part of the Executive Fire Officer program that involve prevention efforts and are readily available for viewing or download from the NFA.
- The National Fire Protection Association (www.nfpa.org)
    - The NFPA maintains a Research Foundation that funds and engages in fire protection research on a wide variety of topics. As a nongovernmental entity, the NFPA must fund its research efforts—so reports usually come with a price attached. However, the quality of the research is high, and most is relevant to a wide variety of fire prevention issues. Most recently, the Research Foundation has been conducting research (as yet unfinished) that deals with the effectiveness of fire code compliance inspections. Such studies are extremely rare, because they are generally not considered a high priority by those outside the fire prevention field, and funding is difficult to obtain.
- The Centers for Disease Control and Prevention (www.cdc.gov)
    - The CDC has conducted numerous studies on fire prevention issues. Most recently, the CDC (together with TriData Corporation, a division of System Planning Corporation of Virginia) has conducted international studies on which programs are producing measurable reductions in fire loss rates in other nations.

- Underwriters Laboratories (www.ul.com)
  - UL conducts a wide variety of research related to fire prevention. As part of their safety standards development, UL routinely develops scientific testing methodologies and applies them to research and testing for fire protection products.
- Factory Mutual (www.fmglobal.com)
  - Factory Mutual is a global insurance organization that conducts research into fire protection systems, but also ventures into human behavior studies.
- Hughes Associates, Inc. (www.haifire.com)
  - Hughes Associates conducts research in fire protection, primarily through its testing laboratories—but ventures into other areas such as a recent study on smoke alarm response: estimation guidelines and tenability issues.

Many other organizations conduct fire prevention research, including the U.S. Department of Energy, the National Science Foundation, the U.S. Department of Housing and Urban Development, the U.S. Department of Transportation, the Bureau of Alcohol, Tobacco and Firearms, and the Federal Bureau of Investigation. The most effective way to identify current research is through an Internet search and by attending trade organizational meetings (such as the annual NFPA meeting), where current research is often a topic of discussion.

# The Value of Research

Innovation sometimes comes by inspiration. The Greek philosopher Plato is credited with saying "Necessity is the mother of invention," and made that statement in about 350 BC. In other words—need is what drives research to find a better way to do things. Trial and error can also be effective, and it is in fact a basic tool of the scientific research method. Establishing a hypothesis, and testing it, could be considered trial and error of a sort.

However, without research, our efforts to improve fire prevention efforts would be severely limited. If we did not study how people reacted in fires, we would not know how to guide them or to create built-in features to protect them. If we did not scientifically examine fire tragedies, we would never learn how hundreds of people can die—or, more importantly, how they could have been saved from the ravages of fire. If we did not research the technology available—or envisioned—for fire protection and prevention, we would not be advancing the field of fire safety.

Research, no matter how rudimentary, is the beginning of making decisions based on evidence rather than conjecture. And when we begin to compare the results of our prevention efforts at local, state, regional, and national levels, we begin to shed more light on the types of prevention programs that produce measurable impacts and outcomes.

## Summary

There are different types of research that advance the field of fire prevention. Pure research involves fact finding and verification of theories for its own sake. Applied research is used to solve a particular problem. Pure research would tell us how water reacts with heat. Applied research would help guide our efforts to create an automatic fire sprinkler system that puts out fires without creating water damage.

Studies on human behavior and technology are applications in research that can help improve our prevention efforts. They are a foundation that can begin to move us in the direction of evidence-based decision making, rather than developing programs based on our personal bias

or conjecture about what might be effective. Fire protection technologies are an important part of our prevention efforts, but local authorities having jurisdiction usually rely on a testing laboratory listing to provide some evidence that products will perform as they are designed and safely.

There are many organizations that conduct fire prevention research in a variety of areas, and an Internet search on the topic or attendance at a trade meeting are the most likely ways to identify current research relevant to the field of fire prevention. The value of all of the various types of research is to improve our efforts, ultimately leading to more effective prevention programs and improved public safety.

## Case Study

One of the fire commissioners (an elected official) for the City of May, Washington, has approached the fire marshal with an idea about prevention efforts in their city. The population of May is about 200,000, and news articles point to the fact that the population figures continue to increase. Vehicle registrations and permits for new housing are all increasing—but a formal census has not been conducted for nearly 10 years. Consequently, accurate information for the demographics of the City of May is lacking. There seem to be an influx of Hispanic and Korean immigrants, based on anecdotal accounts by police and fire encounters. There is a large population of Hispanics in what is known as a low-income part of the City. Fire rates for that area are being tracked through a local version of the National Fire Incident Reporting System, and another software program is used to track patient care information for emergency medical incidents. Both call types appear to

be up for the area as a whole—and for that neighborhood in particular.

Recently, the fire marshal had attended a training at a national fire service meeting where one speaker referred to the fact that many people who are immigrating to the United States—some of them illegally—do not read in their own language or English.

The fire commissioner has an active interest in smoke alarms—but he wants them for his own neighborhood, an upscale part of May where fire incidents are historically low. Medical calls for a portion of the commissioner's neighborhood have begun to increase, especially since an assisted living complex opened. More such complexes are planned for that part of May. They are upscale properties with a view of the river.

The fire marshal faces a challenge in identifying where to promote smoke alarms and how to go about doing so.

# Case Study Review Questions

1. What kind of local research would be helpful in targeting efforts where they will do some good?
2. What kind of challenges does it present to give the fire commissioner data that would indicate the fire problem is much higher in another neighborhood of May?
3. Where would national research be available to help design educational materials for a target population that does not read in any language?
4. What type of search could be done to identify materials already available for the target audience?
5. What is the difference between pure and applied research?
6. Why is human behavior important to fire prevention efforts?
7. What is the value of research for fire prevention efforts?
8. List some agencies or organizations that conduct fire research.
9. How does current research apply to smoke alarms?
10. What fire protection features are based on research?

PEARSON

# myfirekit™

For additional review and practice tests, visit **www.bradybooks.com** and click on MyBradyKit to access book-specific resources for this text! Your instructor may also assign Additional Project work related to topics in this chapter.

Register your access code from the front of our book by going to **www.bradybooks.com** and selecting the mykit links. If the code has already been scratched off, go to **www.bradybooks.com** and follow the MyBradyKit link from there.

# Endnotes

1. Grone, Norman E. (2002). *Fire Protection Engineering Magazine* (issue 16).
2. Wieczoek, C., B. Ditch, and R. Bill (March 2010). *Research Technical Report: Environmental Impact of Automated Fire Sprinklers, Factory Mutual*. Retrieved from http://www.fmglobal.com/assets/pdf/P10062.pdf.

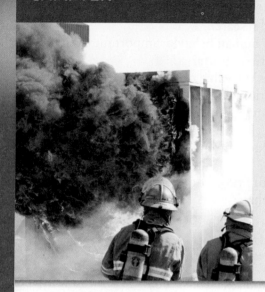

## KEY TERMS

**Environmental analysis** *p. 129*
**Integrated risk management** *p. 138*
**Master planning** *p. 125*
**Organizational culture** *p. 132*

**Potential risks** *p. 125*
**Real risks** *p. 125*
**Regional planning** *p. 139*

**Standard of cover** *p. 126*
**Strategic planning** *p. 127*
**Tactical planning** *p. 134*

## OBJECTIVES

After reading this chapter you should be able to:
- Identify elements of the strategic planning process.
- Assess the differences among master, strategic, and tactical planning.
- List the advantages of planning for prevention programs.
- Identify the major obstacles to planning.

PEARSON
## myfirekit™

For additional review and practice tests, visit **www.bradybooks.com** and click on MyBradyKit to access book-specific resources for this text!

What is the point of planning? We might just as well ask, what is the point of using a map? After all, *any* road will get you there—as long as you do not know where you are going.

The need for planning is so evident that most do not need to be reminded of its importance. They just need to be reminded to take the time to do it right, to revise the plan as it evolves, and to take into account the (always) unforeseen circumstances that arise. It is sometimes incredibly difficult to plan ahead. The demands of everyday work can seem overwhelming enough—especially in times like these, when tremendous pressure is being placed on public officials to do more with less. And despite common phrases such as "work smarter, not harder," the workload just seems to get heavier. For those new to the field, this will not seem quite so important. But ask veterans of the fire service, and especially those in the prevention field, and you will get an earful about how difficult it is just to keep up.

But the fact is that planning is even more important in such trying times. It is a process of creating priorities and targeting our efforts where they will do the most good. It might mean eliminating some programs either because they are not effective, or simply because inadequate resources must be expended where they will maximize our safety efforts. To do less is to miss the bull's-eye of successful planning.

Fire prevention leaders must work in concert with others to develop their plans. We do not operate in a vacuum, and what affects us may affect others; just as what happens to others can have an impact on our prevention programs. An overall budget reduction in the fire department could translate into budget reductions in prevention. At an absolute minimum, the fire chief will have to evaluate the impacts of cuts from emergency operations and compare them with cuts in prevention. Consequently, it is important to plan as an entire department and to figure out how prevention programs fit into a department-wide plan. But our relationships only begin within the fire service.

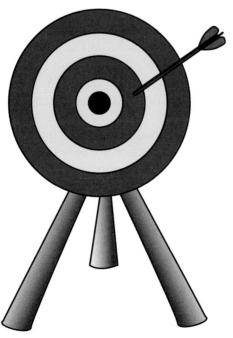

Failing to plan our fire prevention efforts effectively can virtually ensure they will be off target.

We work routinely with the police services whenever we are investigating fires, and in many instances around the nation, when we are collaborating on efforts such as alcohol impact zones or fireworks law enforcement. Alcohol impact zones are set up to reduce alcohol content and sales in areas frequented by transients who often overimbibe and create a demand for both police and fire services. Many jurisdictions across the nation collaborate on fireworks enforcement rules, issuing citations for violators who use fireworks illegally.

We work most frequently with building officials in the area of new construction. The entire plan review and acceptance inspection process (discussed in detail in Chapter 4) joins us to many potential planning partners. It may also bring us in close contact with transportation officials when dealing with fire department access. It will bring us together with water purveyors when we are dealing with fire flow demands and water supply for firefighting purposes, or for fire sprinkler protection.

The plan review process may bring us into potential conflicts with planning and zoning departments—particularly where their goals and ours naturally compete. One such example is in areas of potential wildfire hazards, where planning, zoning, and environmental goals usually revolve around wildlife habitat and water quality protections. We tend to see fuel load where others tend to see habitat or soil protection issues that can affect water runoff and in turn affect water quality.

We often work with school officials when we are trying to reach school-aged children with our prevention and safety messages. Moreover, we are now finding that our fire code regulations, which are designed to get people out of a school quickly in the event of a fire, often conflict with security measures that have been added because of school shootings. When a tragedy such as the Columbine High School mass shooting occurs, school officials, security officers, and police almost immediately lock a school down rather than evacuate children and teachers to where they could make easy targets in the open. Such security goals create a need to coordinate both fire safety evacuation and security plans that may work together.

We may work with private ambulance providers or hospitals on efforts designed to limit falls for the elderly. Those types of emergencies are incredibly expensive for a community, not to mention the debilitating impact they have on the elderly and their quality of life. Falls account for a major portion of injury-related deaths in the population over the age of 65. Calls related to falls account for a high percentage of emergency responses, placing a huge demand on emergency responders.

The number of potential planning partners is limited only by the scope of our prevention mission. Whichever direction we may take to prevent fires or the damage they cause, or if we branch out into injury control issues to reduce our medical call volume, we have potential partners and conflicts. We have potential competitors for available resources as well. Each potential partner may have a stake in the direction our plans may take. So it is critical that we plan if we hope to achieve any measure of success. And it is important that we should not plan in isolation. But how?

What does it mean to plan for life safety and fire prevention? How can we effectively plan when there are so many issues to deal with? There are many overlapping terms regarding planning; and they seem to change with the times—but the basic elements within a planning process really do not change. There are common steps among the many different planning types. Different jurisdictions may favor different types of planning. Some are described here (certainly not all). This chapter focuses most on strategic planning, but also discusses master planning and tactical planning.

# Master Planning

**Master planning** is a term that has been used in the fire service for decades. It is generally attributed to the first executive director of the National Commission on Fire Prevention and Control. That entity grew out of "America Burning," a national report on fire in the United States produced in 1973. That report recommended that "every local fire jurisdiction prepare a master plan designed to meet the community's present and future needs in fire protection." That report, and the resulting interest from the fire service community in the United States, led to the formation of the U.S. Fire Administration and the National Fire Academy.

Since then, master planning, or a derivative of it, has been advocated throughout the fire service and among other decision makers (such as the International City Manager's Association) as a way to prepare the most cost-effective and efficient fire protection programs. There are elements of master planning that relate both to emergency operations and to prevention efforts.

**master planning**
- overall strategic and tactical planning for fire department operations

## STEPS OF MASTER PLANNING

Master planning begins with a survey of resources and personnel of a specific jurisdiction and an analysis of the effectiveness of fire and building codes in that area. Depending on available resources, this part of the planning process can include a detailed analysis of the impact that building and fire codes, and related plan review and inspection services, may have on the fire loss rates of the community. It is a complex set of issues to deal with, one that goes beyond just the number of fire stations, firefighters, and inspection personnel. It may involve an examination of the impacts and outcomes produced (such as changes in the number of hazards identified and abated) that accompany those programs.

The second general step of a master planning process includes a risk assessment of the community, including a short- and long-term analysis of fire prevention and control needs. This step is where the incident and loss data, such as those produced by the National Fire Incident Reporting System would help identify **real risks** and **potential risks**. Real risks are those pointed out by a statistical history of events, such as like the number and particular types of fires (e.g., fires caused by discarded smoking materials) in a community. Potential risks are those that do not have a high incident rate, but would be significant if they occurred (such as a fire in a hospital). The needs associated with each type of risk are evaluated in this phase of a master planning process. This phase also could involve calculating items as specific as fire flow for a particular structure. In some smaller jurisdictions, calculating the fire flow (water supply) necessary for a specific structure, and then extrapolating the number and type of resources necessary to contain a fire in that structure, would be possible. In larger jurisdictions, a risk assessment of this type usually involves some type of modeling and conjecture on a larger scale that would estimate needs for fire protection personnel and resources.

**real risks**
- those pointed out by a statistical history of events

**potential risks**
- those that do not have a high incident rate, but would represent a significant risk (e.g., hospitals) should a fire occur

standard of cover
■ a document, produced through a process similar to master planning, that arrives at local decisions for deployment of emergency response personnel and resources

A more modern version of master planning uses a process from the Center for Public Safety Excellence called the **standard of cover** to determine the needs for fire protection resources. A standard of cover is a document produced through a process similar to master planning. The standard of cover arrives at local decisions for deployment of emergency response personnel and resources. It generally arose from an effort by interests within the National Fire Protection Association to produce a staffing standard for fire protection and emergency response. NFPA 1710 (for predominantly paid fire departments) and NFPA 1720 (for predominantly volunteer fire departments) were produced as an attempt to standardize the fire protection resources needed for any community.

Because there was widespread disagreement within the fire service over the ability to produce a single standard that would fit every jurisdiction, the International Association of Fire Chiefs and the International City Managers Associations joined in a partnership to produce the Council on Fire Accreditation International. That organization later evolved into the Center for Public Safety Excellence (CPSE), which now produces the guide to help local fire departments develop their own standards of cover.

The risk assessment and analysis phase of the master planning process has evolved considerably over the years and is essential for fire prevention programs because of the interrelationship between emergency response and fire prevention methodologies. The woman who is rescued from the fourth floor of an apartment fire lives because there were adequate emergency response resources to fight a fire and to rescue her. Concurrently, a fire sprinkler system required by current codes, and ensured to be in good working condition by aggressive plan review and inspection practices, can limit a fire to a particular area, eliminating the need to rescue that same person.

Both methods of providing for fire safety are part of a master planning process, and as readers will see, part of a modern approach to planning.

## ISO GRADING SCHEDULES

The Insurance Services Office (ISO) provides another view of risk assessments. Commonly, fire departments use some form of ratings system to evaluate their fire protection *and* fire prevention efforts independently. Each area of service receives a "grade" or deficiency points depending on the scope and quality of service, as defined by the ISO grading schedule. For example, a fire department with an inadequate fixed water supply would receive a higher number of deficiency points. That in turn would translate into (potentially) higher insurance premium rates. A department with more frequent fire code inspection cycles would receive lower deficiency points, potentially reducing fire insurance premium rates.

These relationships between the grading schedule and insurance rates are not concrete. Other factors—such as major losses like Hurricane Katrina, where insurance pools must be increased to compensate—can affect insurance premium rates regardless of local services. Economic losses for a specific insurance company can also affect rates, or data on local conditions, such as fire loss statistics that are unusually high (e.g., high arson rates).

The important thing for prevention planners to be aware of is that part of a sound environmental analysis includes looking at the ISO schedules, or an equivalent insurance rating service, for their area. These schedules could be a significant factor in determining how prevention efforts should be designed.

Both emergency response and prevention measures, such as fire sprinklers, are components of a successful fire protection plan. This is a result of a side-by-side fire sprinkler demonstration where firefighters had to put out a fire with and without sprinklers. (*Courtesy of Spokane Fire Department, WA*)

The third step of a master planning process is to plan how the needs of fire prevention and control will be met. No step-by-step process can be identified here—but simply put, how will the perceived gaps in fire prevention or emergency response services be filled?

The fourth step is to estimate the cost of those plans, and how an implementation strategy will be financed. The fifth and final step is to summarize the problems that are anticipated in implementing the plan.

It is important to note that a master plan (or any other plan) should result in an organized document that defines fire protection needs, and a level of protection that will be financed and supported by the community. A part of that means including partners in the planning process, including citizens who will ultimately be served, and assessing the level of risk they are willing to accept. Ultimately, it all boils down to how much citizens are willing to pay, and how much risk they are willing either to manage on their own or to accept in general for the entire community. That is why master planning continues to be so important. Planning with community partners means that anyone with a stake in the outcome of the plan can help identify how it will be achieved.

## STRATEGIC PLANNING

In recent years, many fire departments have begun following planning models related to **strategic planning** that were developed in the private sector. Strategic planning has many elements similar to master planning and ultimately ends up in the same place. It tends to be more focused on higher level strategies, leaving lower-level tactical decisions for another phase or even another planning group.

Many in the fire service are familiar with the traditional form of strategic planning. The broad categories include an analysis of an organization's *strengths, weaknesses, opportunities,* and *threats* (SWOT). A planning process includes examining

**strategic planning**
■ broad directions in a traditionally identified process that includes stakeholder influences on the plan; different from tactical planning

the strengths of a particular organization to meet its mission. Those strengths could include adequate resources, good community support, a good relationship with other providers such as ambulance companies, or a myriad of other factors that mean a fire department is well situated to provide fire protection services.

As the strategic plan is formulated, jurisdictions would also examine weaknesses, such as having inadequate resources, water supply, or a lack of fire inspection services. They usually evaluate opportunities that may exist for improvement, such as strategies designed to obtain more resources, including how they would be gained. In this context, threats that might be identified could include competition from another agency, or political problems that would prevent a fire marshal from obtaining additional funding.

More recently, some users of this planning system have changed the final issue from *threats* to *challenges,* so that the old SWOT analysis becomes SWOC. To help determine strategic directions, a great deal of analysis is required, including an environmental review and a determination of the organizational mission and values from the entire organization. Matching prevention plans within larger issues is always necessary. We need to plan with partners (called stakeholders) and figure out how to achieve prevention goals within a larger fire protection context. Involving stakeholders (those with an interest in the outcome of the plan) is critical. Stakeholders may be internal or external to the organization. In an ideal world, a plan could be developed in relative peace and quiet. But the real world seldom allows sanctuary for planning. Those with an interest in the plan often make loud noises about their stake in the outcome and insist on participation.

Finally, it is not merely a matter of doing things right, but doing the right things. If we do not plan for specific problems and audiences within that larger fire protection context, will miss the mark. There is a link between strategic and tactical planning that will be explained in greater detail in this chapter. The planning process can be extremely messy, but what follows is a specific process developed by Drs. Jim Marshall and Dan O'Toole (now retired) from Portland State University in Portland, Oregon. They spent years developing a strategic planning process specific to fire departments, and this chapter combines elements of strategic planning and tactical planning designed specifically for fire prevention programs.

For the sake of clarity, the planning process for life safety and fire prevention programs will be presented in a fairly sequential order. The process combines a modified strategic planning process with tactical planning derived from the five-step process adapted for use by the U.S. Fire Administration in the early 1980s. Broadly put, the three steps of this version of the planning process include gathering information for planning, planning at the strategic level, and tactical planning.

# Gathering Information for Planning

Gathering information for planning purposes means doing a good deal of research up front. This step is critical, for a variety of reasons. An all-too-common mistake of fire officials is copying a program from elsewhere without any evidence that it is needed or appropriate for their own locale. There are many examples, such as copying a popular puppet fire safety program and then finding out that the children of a particular jurisdiction do not identify with the puppets because their demographic is not represented. Another example: assuming that people like the fire department and will open their doors for a smoke alarm installation program—when just the opposite is true.

Many fire officials assume that they already know their area's needs, rather than digging for information about their fire and injury problems and the people involved in creating them, so that they can use real information in the planning process. The first step in gathering this type of information for planning is to conduct an environmental analysis.

## ENVIRONMENTAL ANALYSIS

The term **environmental analysis** may be misleading if taken out of context. Environmental analysis is not limited to the physical environment, water quality, or the protection of wildlife. It *does* involve reviewing a variety of elements that constitute the "environment" of a governmental entity such as a fire department. The analysis begins with a review of the demographics of the community. Environmental analysis for fire prevention planning is the part of the strategic planning process that provides us with an informational foundation on which to base our decisions. Fire officials must also be careful not to adopt programs from other jurisdictions without first understanding whether those programs will apply in their own jurisdiction. Hence, the need for an environmental analysis and a thorough examination of the audiences of our prevention efforts.

Environmental analysis includes a study of the demographics of the community; an examination of potential risks through a risk assessment; a review of technologies that might affect prevention efforts; a review of the political scene in the community; a look at the future to see where the community and the potential resources are likely to go; and a review of best practices from other jurisdictions to gain some idea of what is working.

**environmental analysis**
■ review of a variety of elements that constitute the "environment" of a governmental entity such as a fire department

The demographics of our nation and our local communities are changing. The methods we use to reach one group may not work with another, so research into who makes up our community is critical.

The demographics of a community are data about the people who live there. Generally stated, community demographic data is an examination of the population's placement, as well as its racial, ethnic, gender, and economic breakdown. Demographics are an important element of planning for fire prevention programs. If a significant part of a community's population is Hispanic, for example, then materials should be designed that would be appealing to that audience. Further demographic research might indicate that a significant portion of another immigrant population is not literate even in their own language. This, too, would affect the design of materials for educational efforts in that community.

National research on demographics and their relation to fire prevention can also be valuable. Studies by the National Fire Protection Association[1] have linked fire-loss rates to economic levels, indicating that the strongest link between fire cause rates and a particular audience is income. That is a stronger connection than race, ethnicity, or gender. (Some fire problems are linked in other specific ways: For example, juvenile-set fires are most likely set by young males.) The point is to gather and analyze the information that can help identify those types of targets and links. Whether at a national, statewide or local level, the environmental review should include research into the demographics of our community and their relationship to our fire problems.

But an environmental analysis also includes a look at the political landscape. Information about who makes decisions and how these decision makers are influenced is critical for fire officials who are trying to get programs adopted and funded. It will also help officials align their strategic directions with the needs or desires of local political and community leaders.

A review of the statistical loss history is also part of the environmental analysis. Knowing the answers to the "who, what, when, where, and how" questions about fire incidents will lead to more specifically targeted programs. In other words, where are fire problems occurring, and how are they happening? Who is responsible, and what specific sequence of events led to the occurrence? This kind of information is generally derived from incident reports and will be a significant factor when designing prevention programs, because it helps us to target our efforts where they will do the most good.

As previously mentioned, statistical loss records are only part of the risk picture for a local community. There are many more potential risks that pose a threat to prevention efforts but lead to few incidents. For example, a local nursing home may not have a history of fires, but the fact that one *could* occur demands extra attention from local planners. The risk posed by having so many people who would be difficult to move in a fire congregated in one place is a significant part of the environmental analysis. This type of risk assessment is exactly the same as that mentioned for master planning earlier in the chapter. It should include a search for different types of hazards, such as the wildfire fuel load in an urban community or a listing of building sites that make use of large quantities of hazardous materials. The risk assessment, therefore, is an analysis of the sites and hazards within a community that represent a potential problem but experience relatively few incidents.

A review of the technological aspect of prevention planning is also part of the environmental analysis. Technology plays an increasingly important role in strategic and tactical planning. Knowing what technology is available to assist prevention efforts, such as computer programs for fire modeling or handheld computers for inspections, can change a plan significantly. Fire sprinklers and smoke alarms, cooking safety devices, and other technologies can be the underpinning of a successful prevention program. Having some idea of what will be available in the future is also important. Thus, comparing "what is" to "what will be" is a valuable part of the environmental analysis.

Futures analysis should be applied to each area of the environmental analysis. The fact that most populations are aging is significant. The changing economic picture for a community can have tremendous implications for planning. An environmental analysis not only looks at the "now," but projects into the future so that plans can be prepared in advance.

Knowing the political landscape and where it is headed is a vital factor in the planning process. Forecasting future political trends is also risky. But we should not ignore the power of political influences to shape government services. Researching those political forces in the community does *not* mean taking sides or gambling on an influential winner to improve the position of our efforts. It does mean paying attention to how those might affect our efforts. For example, recognizing that a particular voting constituency is on the rise, and that there is a relationship between that population and our fire problem, offers us an opportunity to position our programs where they could gain political support.

Another common element to the environmental analysis is looking at best practices from elsewhere. Decision makers should be mindful that not everything works everywhere. However, a review of other programs (including those available commercially) can be helpful when it comes to making decisions about buying or developing a prevention program. One important element of a "best practice" review is the criteria under which a particular program may be considered as a "best practice." At a minimum, there should be evidence that the program is producing results—though the claims behind declarations of success should be examined carefully. A review of best practices from other jurisdictions can be valuable, if we are diligent in our research and understand how the programs work—and what results are actually being produced. More is said about this topic in Chapter 12.

In all cases, people responsible for planning may want to seek professional assistance, especially for the futures portion of the environmental analysis. We do not always have the resources and help we need; in those cases we must do the best we can on our own. But often local colleges and universities are an excellent resource when it comes to assisting fire departments and prevention managers with an environmental analysis.

## VALUES AND MISSION OF THE ORGANIZATION

In addition to the environmental analysis, the values and mission of the organization play an important part in the first phase of the planning process. We know that we should not plan in a vacuum. A review of these factors may be one of the most critical parts of the information-gathering process. Local fire prevention planners should try to identify and understand the values of the organization for which they are planning. For example, knowing whether the organization tolerates change and how much of an issue that change will be can affect (even derail) any plan. A review of the current mission statement for the organization, if it has one, can reveal whether strategic opportunities in prevention will match the organizational mission or whether they might conflict instead.

The mission statement is really a lesser factor in determining how prevention programs fit into the larger organizational directions. Often, the values, or organizational culture, of an organization are far stronger than any written document. **Organizational culture** refers to an organization's unwritten values and norms of behavior that are the foundation of its operation. An example would be the team aspect of firefighting. Because firefighters must function as a team, the organizational value placed on teamwork and a collaborative attitude is extremely high. Anyone who did not exhibit those team attributes or attitudes would tend to be ostracized by their peers for not being a team player, even though they were not necessarily in violation of any specific rules.

The key to understanding importance of organizational values in this context is in figuring out where fire prevention efforts fit—or how they can be made to fit—in an organization.

The National Fallen Firefighters Foundation (NFFF) has undertaken efforts to influence the collective organizational culture of the fire service to be more safety prone. For some who are attracted to the fire service for the thrill of risk taking, it is counter to their values to be more safety conscious—but a deliberate strategy to help change those values has been begun by the NFFF. Most recently, that organization and the Institution of Fire Engineers U.S. Branch have begun to collaborate on a campaign to influence the organizational culture of the fire service and to improve the standing that fire prevention efforts have within the fire service.

Such efforts cannot be identified, let alone begun, unless we are looking at the organizational mission and values as they relate to our efforts. Many fire departments experienced this kind of evolutionary value change when they began responding to emergency medical calls within their community. There was tremendous resistance to taking on medical calls and getting away from the core mission of fighting fires. Many fire departments still do resist the concept, but with most experiencing 70% or more of their call volume as emergency medical related, they have been forced to change their operations and their values about what is important to the organization. Similar changes are occurring for prevention efforts, as many decision makers ponder the importance of prevention to their overall fire protection goals. They cannot ignore the cost effectiveness of many prevention efforts—including fixed fire protection features—and are adjusting their strategic plans accordingly.

**organizational culture**
■ unwritten values and norms of behavior for an organization that are the foundation of its operation

# Strategic Planning

The traditional strategic planning process includes an analysis of the strengths, weaknesses, opportunities, and threats/challenges facing an organization [SWOT(C)]. Placed against the backdrop of the environmental analysis, the SWOT(C) analysis

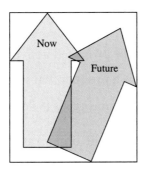

Strategic planning must take into account what is being done currently, and deal with the changing directions required by factors such as changing demographics, changes in fire incident rates for particular types of property, or political changes that reduce overall resources.

can reveal potential political or other environmental challenges and opportunities an organization must face if it wishes to make progress.

Professors Dan O'Toole and Jim Marshall (retired) of Portland State University in Oregon have developed a hybrid of the strategic planning process that has been used successfully by many fire departments. It includes an environmental analysis, but also uses a group-planning model to compare existing programs with strategic goals in order to create a "road map" for a fire department.

As described by Professors O'Toole and Marshall, strategic planning is a

long-term, coordinated, comprehensive plan for meeting goals. It is based on explicit assumptions as well as an investigation of past, present, and future conditions. The plan covers 3 to 10 years, depending on the organization and its environment, and focuses on the organization's target populations, its service area, potential technological advancements, and service modifications required to meet needs. It contains a strategy and a set of tactics for adapting the organization, over time, to its future environment.

Generally stated, the planning process combines the environmental analysis, the futures research, and the values of the organization to produce broad directions for the future. The group process of developing the directions includes a cross-section of internal and external stakeholders who come together for that purpose. Eleven different aspects should be reviewed during the planning process:

1. The time frame of the plan
2. The current organizational mission and goals
3. Major "levers" or "driving forces" considered in the environmental analysis (e.g., the economy, demographics, and the political scene) that shape the future of the organization
4. Opportunities and problems revealed in the environmental analysis
5. Proposed changes (if any) to the organizational mission and/or goals
6. Organizational activities that reinforce the revised mission and/or goals
7. New organizational activities needed to meet the mission and/or goals
8. Opportunities to take advantage of—including the resources necessary to do so
9. Problems to manage or solve in order to implement changes

10. Proposed efforts to change or reinforce the broad direction along which the organization is moving
11. A chronological sequence of proposed activities

The result of this process is a product that includes the following components:

- The organizational mission (and goals if desired)
- The organizational values
- The assumptions made about the organization's future external and internal environments
- A set of strategic directions for the organization to move into the future (more, same, less emphasis on present activities; activities to eliminate, new activities to initiate)
- A set of tactical plans to get the organization moving along the strategic directions

In the final element of the planning process (Element 11), the organization turns to the tactical level of planning. However, the broad "strategic" directions are usually described in terms that require (a) more, the same, or less emphasis on what is currently being done; (b) the elimination of activities that are no longer important to the organization's mission; and (c) new activities made necessary or desirable by emerging changes in the environment.

For example, a strategic direction might include a statement to put more emphasis on taking advantage of (or developing) community coalitions to gain support and resources for a prevention effort. Using this process, the assumption would be that the local planners had already conducted an environmental analysis and decided that given the local political scene, a coalition might be more effective than the fire department going it alone. The review of current goals and objectives could yield a realization that more emphasis is needed on some activities, or new activities (and possibly resources) would be needed if the organization were to take advantage of this opportunity.

At the tactical level, it might become evident that (given finite resources), something else might have to be dropped in order to take advantage of this coalition.

The tactical level is where prevention planning becomes far more detailed. A specific set of factors should be reviewed during the tactical planning phase.

# Tactical Planning

**tactical planning**
■ more detailed than strategic planning; provides a "how to" approach once the broad directions are identified

**Tactical planning** of prevention programs assumes that the environmental analysis has already occurred and that broad strategies are identified. The tactics are derived from the broad directions developed for the organization, but are in fact targeted to specific problems. If the strategic direction were to put more emphasis on fire and life safety efforts, the tactical planning would identify how that would occur. A careful review of the historical loss data, as well as the listing of potential problems for the community, would be required.

In broad steps, the tactical portion of the planning process is based on the five-step process adapted for use by the U.S. Fire Administration in the early 1980s.

The specific steps include identification, selection, design, implementation, and evaluation.

## IDENTIFICATION

In the identification phase, decision makers closely examine the data to create a series of problem scenarios for which specific program efforts may be targeted. During this step, historical data are analyzed to determine high-risk locations, times, victims, attitudes, and behaviors for a given community. This portion of the plan might reveal a problem statement like this:

> Fires are occurring at a much higher rate in the northeast portion of the community. They are occurring most often on Friday and Saturday evenings, in one- and two-family dwellings. They involve elderly residents who are smoking unsafely, which often include contributing factors such as alcohol or medication use.

If data is insufficient for this kind of detail, then planners must make some assumptions based on what they know or can safely presume. National studies (for example, from the National Fire Protection Association) about the relationship between causal factors and emergency incidents are available. However, the best way to identify real problems and create solutions that work is to look at local information.

The identification phase also allows planners to create scenarios that deal with potential problems. Another scenario might read like this:

> The wildland/urban interface problem for this community is seasonal in nature, and occurrence levels are low. However, the potential economic and life losses for the community if a fire occurred would be catastrophic. A review of the high-hazard areas of the community has been developed using topographical maps, fuel load analysis in specific areas, and a physical review by emergency operations companies. The plan identifies seven areas within the city limits that would be viewed as high hazard during dry conditions.

It is up to local decision makers to decide which types of scenarios are most important for their community, given their own history and a review of the potential problems they may face.

## SELECTION

Under this portion of the tactical plan, several objectives might be outlined from which a few would be selected for action. Obviously, a review of resources is necessary to determine how much of the "apple" a community can afford to bite into. Equally obvious is the need to conduct some kind of cost/benefit analysis to help select the objectives that will yield the best results. And particularly for public education efforts, this is the phase where audience factors should be examined even more closely. Doing so will help in designing programs later in the tactical planning process.

If, for example, the targeted population is elderly, some thought should be given as to how they will be motivated to change their behaviors. Planners must realize that each individual audience may be motivated by different factors. Some may be more motivated by outside influences such as their church than by any official government entity.

Once these factors have been reviewed, planners should select specific objectives that support the broader goals of the organization. The next phase of tactical planning involves development and design.

## DESIGN

Once objectives have been selected, development of specific programs may begin. Planners should review other sources of information to determine whether their programs need to be "reinvented" or whether they can be purchased or adapted from elsewhere. There is an obvious relationship between the program effort and the problem identified. For example, most communities understand the value of fire code inspections. Inspections are conducted because the failure to do so elsewhere has produced some catastrophic incidents. A variety of inspection programs already exist, and the design portion of the plan usually involves the tactical considerations of implementation.

However, if another problem is identified, such as fires in residential units, an inspection program will have limited value. This is true because most inspection authorities do not have jurisdiction for one- and two-family dwellings. Under these circumstances, the program design must take an educational approach and examine the specific audience and type of problem to be effective.

The elements to consider when designing a specific prevention program for educational purposes includes the following:

- The message content should be relevant, positive, and informative and tell people what *to* do, rather than what *not* to do. For example, telling children to give matches and lighters to adults is more effective than telling them not to play with them.
- The message format (delivery media) should be appealing to the target audience. For example, puppet shows are great for kids, but seldom work on adult audiences. Video works well with just about any audience, but for many ethnic groups the wrong kind of narrator or actor can limit its acceptance—just as the right kind of format and narrator can appeal to a given audience.
- The message time and locations should match the problems identified. Brochures developed for a wildland/urban interface problem will have limited appeal in the wet winter months. Likewise, public service announcements about Christmas tree safety will not likely be given much attention in the summer. Also, placing brochures where they are unlikely to be viewed by the target audience is a waste of time and money.

For development of any program, it is almost always a good idea to test it before full implementation begins. Using a focus group made up of members of the target group is an inexpensive, effective way to get feedback about a program. A focus group can uncover problems while they are small enough to be managed—and before great expense has been undertaken.

## IMPLEMENTATION

Implementation strategies are dependent on the environment. The environmental analysis, especially the review of the political environment, will have a major impact on implementation decisions. In addition, the review of potential audiences will determine how a program is implemented. However, some broad topical areas exist for any implementation plan. They include:

- **Producing or Procuring Materials/Resources.** This is where local planners need to make the determination to develop something specific or to buy it "off the shelf" to save development time. Time is taken at this point to determine what resources are available and what is needed.
- **Funding.** At this point, planners usually take into account whether alternate funding of some type is needed or whether other programs will be scrapped in favor of a new effort.
- **Methods of Distribution.** If, for example, brochures are developed or purchased to get a specific message to the public, planners must determine how they will be distributed.
- **Finding and Training Personnel.** The personnel requirements (paid or volunteer) should be identified. Planners should also account for the necessary training time and expense to the program.
- **Scheduling.** Decision makers should determine what the frequency of the program will be and how it will be scheduled. This step has obvious ramifications for the number of personnel needed to adequately perform assigned tasks.
- **Obtaining Cooperation and Support.** If not addressed earlier in the process, this step of tactical planning will provide planners with the opportunity to *plan* how support and cooperation will be obtained. The tie to the environmental analysis is important, because it should help define who the proper decision makers are and how the support of key stakeholders will be obtained.
- **Monitoring and Modifying the Program.** Planners should prepare for the unexpected and arrange for checkpoints (or milestones) to see how the program is performing and what changes may be necessary.

It is very important that planners understand the implementation aspect of planning. All of the steps of a plan are developed *before* the plan is actually implemented. This means that the implementation and evaluation steps are planned before any activity occurs. Doing so helps to ensure that all the aspects of a program have been considered before a program is started—helping to reduce problems before they occur.

## EVALUATION

Evaluation of life safety and fire prevention programs is covered more thoroughly in Chapter 12. Evaluation involves documenting the changes that may be attributed to the prevention program, such as cognitive educational changes, changing fire risk by removing hazards or installing smoke alarms, or changing behaviors enough to lower fire incident rates over time. Evaluation does not just refer to the results achieved in public safety, but to their cost effectiveness as well. In today's difficult economic environment, people want to know what their tax dollars are buying. Consequently, policy makers usually put pressure on fire prevention planners and managers to demonstrate the impact of prevention efforts and the costs associated with those impacts. In broad terms, they wish to know what results will be achieved, how many resources will be required, and how efficiently they will be managed. They will also want to compare those results with other efforts to determine their own funding priorities.

### FIVE-STEP PLANNING PROCESS

1. Identification: analyzing the data and other information to identify who, what, when, where and how fires (or injuries) are occurring. This step helps to target prevention solutions toward specific problems.
2. Selection: determining which of the problem areas will be addressed – and selecting specific prevention strategies to meet them.
3. Design: designing prevention programs and materials to correct the specific problem – for the specific intended audience. This step helps to match materials and messages to audiences so that they will be likely to get the message – and act on it.
4. Implementation: creating an implementation strategy, including production, purchase and distribution of materials.
5. Evaluation: creating an evaluation plan that monitors the program for progress, allows mid-course corrections, and establishes evaluation measures that can demonstrate the impacts and outcomes a program achieves.

*Note:* all five steps are planned out BEFORE the program/project is actually implemented.

# Integrated Risk Management

The concept of integrated risk management is not new, and it is not limited to the fire service. A quick Internet search will turn up a number of resources available and several different perspectives on the topic. In recent years, the concept of **integrated risk management** (IRM) has been gaining ground in the United States as a new way of looking at planning for fire protection.

**integrated risk management**
- planning for a combination of risks and a balance of methodologies (e.g., emergency response and prevention) to deal with them

Integrated risk management in this context (fire protection) refers to a comprehensive way of planning. The planning elements from IRM are very similar to master planning and strategic planning in that the foundation is conducting a comprehensive risk assessment of a given community. Integrated risk management is then planning for a combination of risks, and a balance of methodologies (e.g., emergency response and prevention) to deal with that risk.

Relative to the fire service, the issue of integrated risk management began in the United Kingdom, where the concept of standards of cover for deployment of emergency response originated. IRM first assesses risk levels, in both potential and statistical terms. As a result of the emphasis on statistical value, this approach leans toward preventive efforts and residential occupancies, where most fire deaths occur.

Paul Young, the former Chief of the London Fire Brigade, attributes the beginning of Integrated Risk Management Plans to Professor Sir George Bain in a report on the future of the fire service and promoting risk reduction. A white paper was produced in 2003 (available on the Risk Institute Web site), along with legislation that required an integrated risk management approach to fire protection services. That approach included assessing local risks, organizing prevention strategies, and allocating response resources according to that risk analysis.[2]

The article goes on to explain that each local fire brigade will have responsibility for reducing the number of fires and other emergency incidents, reducing life loss and injuries, safeguarding the environment and heritage (buildings and natural) and providing value for the money spent on fire protection.

This process of creating an integrated risk management plan marked a decisive departure from previous efforts in the United Kingdom to develop standards of fire protection cover, efforts that focused much more heavily on emergency response.

## RISK ASSESSMENT METHODOLOGY

According to the article, the emphasis on integrated risk management led to the development of a common risk assessment software program funded by the central government for local use. Most departments or fire brigades have adopted the risk assessment method over the past few years. The program allows local departments to assess the distribution of risk in their areas, identify the resources necessary to deal with that risk, and estimate the consequences of various scenarios of protection.

This method of risk management allows the local department to take into account the costs and benefits associated with particular patterns of local emergency cover. The reported weakness of the model is in predicting the consequences of risk reduction measures such as community fire safety education or installing smoke alarms. However, reports on global concepts in fire protection from Tri-Data Corporation and the Centers for Disease Control and Prevention indicate overall reductions in fire death rates by as much as 40 percent using this integrated risk management approach.

The concept of integrated risk management will become increasingly important in the United States as the call for this type of planning increases. Already there are indications that the International Association of Fire Chiefs and the Center for Public Safety Excellence are looking at this type of planning to a much larger extent, to enhance the planning currently undertaken as part of the standard of cover planning process and the accreditation process for fire departments through the CPSE.

# Regional Planning

**Regional planning** is derived from the need to plan beyond the borders of our own jurisdictions with neighbors, and with partners whose jurisdictions have boundaries wider than our own.

Mutual aid may be provided from another fire department whose firefighters may be operating out of another jurisdiction's fire stations. They will be facing the same risks as those permanently providing service for that area, but will be less prepared to deal with it.

Often, jurisdictional boundaries are exceeded by service providers who may be partners in our prevention efforts. An excellent example is the Bureau of Alcohol, Tobacco, Firearms and Explosives. This federal agency maintains regional fire investigation teams that can be extremely valuable on complex fires, so planning without taking them into account would be unwise. There are regional services of many types, from hazardous material response to medical providers including air medical units, where regional planning comes into play. Regional fire prevention efforts are also becoming more prevalent. That is especially true where training

**regional planning**
■ a collaborative process of planning for mutual benefit with those who may provide us resources in time of need, or who may need help themselves

resources are more strained because of tough economic realities. Combining efforts and resources to provide training that may be common to regional fire prevention interests may be one aspect of regional planning. Another is for public education programs that span jurisdictional borders.

Any public education effort that utilizes media—and there are many examples, such as cable television, broadcast television or radio, newspapers, and online news outlets—will transcend jurisdictional boundaries and almost certainly require cooperation. For example, a public education campaign with a radio public service announcement (or a paid advertisement, which is now more frequent) would be competing for resources from other departments with similar interests. Hearing the name of one jurisdiction in a region provides a little local pride in that organization—but often spurs other jurisdictions to do likewise, and the potential for competition increases. Anytime regional public education efforts can be organized, while meeting local needs for getting organizational identity out to taxpaying (and voting) constituents, a bonus of collaboration and increased resources can occur.

There is no set way to provide for regional planning. The many potential partners and their priorities, personalities, and political problems make regional planning extremely difficult. Ultimately it will come down to keeping the planning process simple, identifying a few issues for which mutual cooperation may be achieved, and focusing on the relationships that keep regional planning and cooperation working.

# Obstacles to Planning

There are a number of obstacles to successful planning. The topic deserves special attention not because it is lengthy, but because it is important. A common axiom is that "those who fail to plan, plan to fail." But as already mentioned, the pressure to conduct day-to-day business is probably the biggest obstacle to planning. When a local developer with close ties to community political leaders is clamoring for plan review and inspection services, the pressure can be intense to perform that service in a timely manner (often meaning immediately). When a fire occurs in which a fire death results, and the local media is calling for information about the cause, the need to serve their information needs will not go away until those needs are met. Add an element of suspected arson, and the investigation itself, combined with the heightened media and public (and political) interest, will add a tremendous amount of pressure to perform and to do so quickly.

Conducting a planning process in the face of these kinds of urgent priorities would be practically impossible. Yet situations like these are common in today's fire service because of heavy workload demands and increased public and media attention.

Another common obstacle is a lack of resources or expertise. Most prevention practitioners and even managers are hired for their technical knowledge, not their expertise in planning. So obtaining expertise and assistance can mean the difference between successful planning and simple guesswork. As previously mentioned, local colleges and universities may have students who can help with planning. Local service clubs may be another option, and in some cases a local chamber of commerce may provide business planning assistance that could be translated into strategic planning for fire prevention programs. Local prevention managers may be surprised to find that the chamber of commerce has an active interest in fire prevention efforts and in ensuring that their government services are well planned—executed and run in a businesslike manner. And there may be some added benefit to having business people involved in planning. Those involved in development of a plan become stakeholders in the success of the outcome. As local business people become more aware of the problems associated with prevention efforts, they may in fact add resources of their own to solving some of the problems.

There are a number of resources available on the state and national level to aid in planning efforts. Local planners should consult with their state fire marshal's office to see what kinds of planning tools (software, training, planning guides) might be available. The U.S. Fire Administration has also produced a number of packages over the years that teach local prevention managers how to plan. In particular, the search for and use of data may be supported through the national fire data center at the USFA.

The National Fire Protection Association also produces numerous reports that help identify where and how fire problems occur. They have collaborated on a number of specific reports with the USFA on smoke alarms, kitchen fires, candle fires, and more, whose insights can be extrapolated into a local planning process where resources to research local data do not exist.

For wildfire hazards, there is some assistance and planning guidance available at the National Firewise Communities Web site. The purpose of this multiagency alliance is to involve all stakeholders—home owners, planners, developers, and so forth—in the effort to reduce wildfires before they happen.

Finally, there are courses at the National Fire Academy, and often at fire service trade conferences, that teach local practitioners and managers how to plan.

Resources are available, and there is always a way around the obstacles to planning, even if the final result is not scientifically sophisticated. Doing nothing and guessing about what we should do—or expecting what may have worked elsewhere will work in our own jurisdiction—could be a waste of energy and resources.

## Summary

Planning for prevention programs is both strategic and tactical. Master planning elements are inherent in the strategic planning process and are founded by a thorough risk assessment of a given jurisdiction. The risk assessment should examine both real (statistical) and potential (low-incident but costly) risks for that community.

Planners must look at the broader environment (political, demographic, etc.) and determine how a plan will fit into a larger community vision. If possible, planning must involve the participation of key stakeholders in the development of the plan. Planning generally examines the strengths, weaknesses, opportunities, and threats (challenges) of the organization. However, it does so in relation to the current programs being offered and requires determining where scarce resources will be placed. Finally, it translates general strategic directions to a tactical, working level where details are worked out for implementation.

Integrated risk management is another comprehensive way of looking at planning for fire protection, in which where prevention efforts become a key factor in meeting a community's fire risks and doing something to mitigate them. Regional planning will become important as we realize that not all the resources available to mitigate our community's fire problem belong to us—that in fact using some resources involves cooperating with other agencies outside our jurisdiction.

Finally, overcoming obstacles to planning is critical if we hope to steer our prevention efforts where they are likely to do the most good.

## Case Study

### Strategic Plan for Starchville County Fire District

Starchville County Fire District has six fire stations covering 200 square miles. It serves a population of about 35,000 people, utilizing 40 career and 65 volunteer firefighters. The district offers a full spectrum of fire protection and prevention services, including code enforcement, plan review, fire investigation, and public education programs. Their strategic plan encompasses all the fire department services and serves as the basis for determining whether fire prevention programs are meeting broader department and community goals. Their general process follows the SWOT orientation toward strategic planning.

The plan includes an extensive environmental analysis, including a detailed historical study of incidents occurring within the district using data that spans 12 years.

Generally, their plan answers the questions: where are we now; where are we going; how are we going to get there; and how are we going to measure success? It stipulates that the mission of Starchville County Fire District is to protect life and property through public education, fire prevention, and emergency response services.

Their vision statement includes the following elements:

- Provide high-quality customer service
- Leader in local and regional training
- Committed to personnel safety in all activities
- Committed to a combined paid/volunteer delivery system
- Fast and effective emergency response
- Responsive to social, economic, community, and technological change
- Committed to organizational and personal education and training
- Highly functional facilities, apparatus, and equipment

Their SWOT analysis revealed the following:

- *Strengths:* very experienced and knowledgeable work force; highly integrated volunteer firefighter program; positive reputation and working relationships with other agencies; functional and up-to-date facilities, apparatus, and equipment; positive labor management relationships; capability to meet diverse emergency response situations; organizational and personal emphasis on training; high visibility and participation in community activities
- *Weaknesses:* work-flow coordination between shifts, divisions, and stations; career development and promotional training programs; insufficient interagency training, knowledge of new high-tech businesses moving into the district; maintenance of on-site water storage systems and wildfire hazard maps
- *Opportunities:* fire protection contract with City of Hughes; updated computer and communications equipment; continued emphasis on community involvement; new computer-aided dispatching and records management systems
- *Threats:* new restrictive federal and state mandates and/or other national standards; restricted operational/capital funding; turnover of paid personnel; reduced ability to maintain response times with increasing call volumes; residential development in wildfire hazard areas

From the analysis, several directions and operational objectives were developed. Only a few are listed here:

- Ensuring a high level of disaster preparedness for floods, earthquakes, wildfires, and other predictable disasters
- Ensuring a high level of public knowledge of district services and fire safety practices
- Maximizing administrative efficiency and effectiveness

From these general directions, a number of tactical objectives were established. Among them were several related to fire and life safety:

- Provide wildfire safety information to home-owners in wildfire hazard areas
- Publish three educational newsletters annually
- Investigate options to support the installation of sprinkler systems in new and existing residential developments
- Review the district service fee schedules and procedures

Responsible persons were assigned to the many tactical objectives derived from the strategic directions. Timelines and reporting procedures were developed for each tactical objective. The charts shown here are an example of the implementation and monitoring system employed by Starchville County Fire District.

Recently, the fire district board of directors asked for more detailed information about the prevention efforts identified in the plan. They wanted to know how much the newsletters cost, why they were included in the plan, and whether they were producing any results in the community. They also asked what benefit the urban/wildland interface educational efforts were producing for their community.

# City of Starchville Performance Reports

## 2008 Fire Department Performance Snapshot

### Fire Chief: Mike DiNenno

<u>Mission:</u> We provide highly trained professionals, well-equipped to respond effectively to the education, prevention, and emergency response needs of our community.

### 1. Synopsis of 2008 Performance

    a. SFD had improved response times for urban and rural

    b. Starchville Fire department Standard of Cover is still in progress

    c. Completed data management overview and are working on source all data points

    d. Business Plan-Complete

    e. Acquired land and permit for Station 810

### 2. Key Performance Measures

White: No goal; Green: Fully met the goal; Yellow: Missed the goal, but are close OR provided limited service; Red: Clearly missed the goal OR a very bad trend; **FPY** = From Previous Year; ⬆**up = trend better** ⬇**is trend worse**; ≤ is "less than or equal"; ≥ means "greater or equal"

| Performance Measures & Outcomes | Current Goal | 2006 | 2007 | 2008 | Trend | Analysis |
|---|---|---|---|---|---|---|
| **We respond quickly to emergencies** | | | | | | |
| Total calls for service - all types | decrease FPY | 21,203 | 21,326 | 23,184 | | |
| Urban/City response (81.5% of call volume) arrive within 5 minutes | 90% | 57% | 59% | 61% | | |
| Suburban response (5.9% of call volume) arrive within 6 minutes | 90% | 55% | 54% | 52% | | |
| Rural response (2.9% of call volume) arrive within 8 minutes | 90% | 71% | 69% | 73% | | |
| False Alarms - False alarms as a percent of total fire/EMS calls | < 5% | 3.8% | 3.4% | 4.7% | | |
| **Our medical response is effective (EMS)** | | | | | | |
| Cardiac arrest survival rate (calls where revival attempted) | increase FPY | 13.6% | 26.5% | 57% | 👍 | |
| Advanced Life Support: percent of calls to nursing, assisted living & medical facilities[2] | decrease FPY | 14.3% | 8.3% | 7.9% | | |
| Advanced Life Support (Paramedic) average on-scene time | decrease FPY | 17:06 | 18:00 | 16:03 | 👎 | |
| **We minimize the effect of fires (Fire Service)** | | | | | | |
| Percent of fires confined to room of origin | increase FPY | 49% | 50% | 40% | | |
| Average dollar loss due to structure fires | decrease FPY | $15,523 | $12,585 | $41,462 | | |
| Average total fire response on-scene time | decrease FPY | 44:53 | 55:11 | 47:11 | 👎 | |
| **We prevent fires (Fire Prevention)** | | | | | | |
| Determine cause of fires | > 80% | 78% | 90% | 89% | | |
| Conduct safety presentations | > 200 | 241 | 186 | 183 | | |
| Number of effective fire inspection programs | > 5,000 | 5,669 | 5,780 | 6,006 | | |

Each graphic tells part of the story of the environmental analysis conducted for Starchville County Fire District. Mostly, they center on the statistical fire incident rates. Which parts of the environmental analysis are missing?

*(continued)*

| We respond to unusual situations (Special Operations) | | | | | | |
|---|---|---|---|---|---|---|
| Number of hazardous materials emergency responses | decrease FPY | 118 | 124 | 130 | | |
| Number of marine emergency responses | decrease FPY | 29 | 31 | 30 | | |
| Number of technical rescue emergency responses | decrease FPY | 19 | 20 | 30 | | |
| **We provide highly trained first responders (Training)** | | | | | | |
| Fire academy average cost per recruit | < $16,000 | $11,430 | $14,574 | n/a | | |
| Multi-company training drills per shift | ≥ 4 | 6 | 4 | 12 | | |
| Percent of FOTEP attendance (all shifts) | 100% | 98% | 98% | 100% | | |
| **We use resources responsibly** | | | | | | |
| Overtime as a percent of personnel costs | < 5% | 9.12% | 8.45% | 8.66% | | |
| Overtime costs | decrease FPY | 1,628,174 | 1,764,122 | 1,797,317 | | |
| Change in overtime use from previous year | decrease FPY | 3.66% | -.68% | .21% | | |
| Grant and federal fund acquisition | increase FPY | $593,350 | $210,251 | $316,391 | | |

Note 1: 88% of calls received are medical, but many are canceled; 70% of calls we arrive on scene are medical
Note 2: Significant effort made to reduce this call type, which is not an emergency call

| 3. Sample cost to provide services | 2007 | 2008 |
|---|---|---|
| a. Average cost per Fire/EMS response* | $1,409 | $1,251 |

*Total expenditures/total calls

### 4. Major issues to address 2009–2010

a. Build a LEED certified Fire Station by January 2010
b. Initiate the first Firefighter I/II certification program through a partnership with Clark College
c. Hire and train 24 new firefighters in 2009
d. Continue to monitor and research new ways to control overtime
e. Research new revenue sources in 2009
f. Implement Accounts Receivable hazmat fee using firehouse
g. Continue to work to integrate software to improve data collection and accuracy
h. Research consolidating SFD training division with other Clark County Fire agency training divisions
i. Continue to work at capturing cost at each station
j. Begin implementation integrated risk management
k. Complete two Citizen Academy's in 2009

### 5. Department Contacts

Name:          Email:                    Telephone:
Chief Mike DiNenno

(A)

*(continued)*

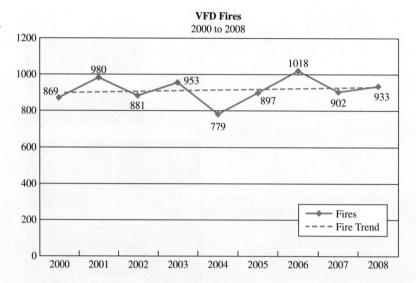

**VFD Fires**
2000 to 2008

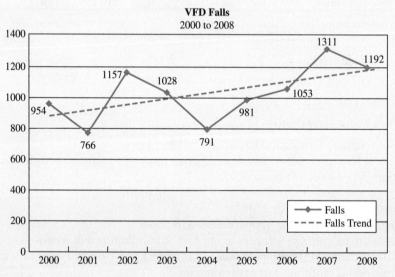

**VFD Falls**
2000 to 2008

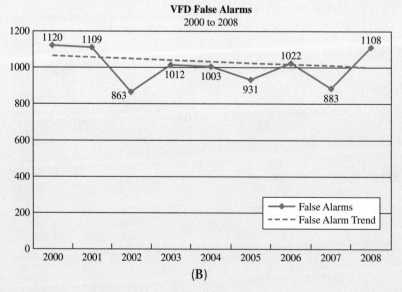

**VFD False Alarms**
2000 to 2008

**(B)**

# Case Study Review Questions

1. How will an examination of the strengths and weaknesses of any organization affect its ability to carry out a plan?
2. Describe the relationship between organizational directions and fire prevention activities using examples from the case study.
3. Given the case study, explain how external environmental factors might affect the implementation of some objectives.
4. Explain how one might identify the data to answer the questions produced by the board of directors.
5. Using the case study, describe how the prevention strategies did or did not relate to the environmental analysis, including the data provided; describe what other tactical prevention objectives might be developed from the same data. Focus most specifically on the mailing of newsletters, and what other options for education or other prevention efforts might be available.
6. List the elements of strategic planning.
7. Why is the use of stakeholders during planning so important?
8. What are some differences between strategic and tactical planning?
9. Why is planning so important?
10. What are some common obstacles to planning?

PEARSON

## myfirekit

For additional review and practice tests, visit **www.bradybooks.com** and click on MyBradyKit to access book-specific resources for this text! Your instructor may also assign Additional Project work related to topics in this chapter.

Register your access code from the front of our book by going to **www.bradybooks.com** and selecting the mykit links. If the code has already been scratched off, go to **www.bradybooks.com** and follow the MyBradyKit link from there.

# Endnotes

1. Hall, John R., Jr., *U.S. Fire Death Patterns by State*. Quincy, MA: National Fire Protection Association, Fire Analysis and Research Division.
2. Young, P., (2003). Integrated Risk Management and Firefighter Safety in the United Kingdom Fire and Rescue Service. Retrieved from https://www.riskinstitute.org/peri/index.php?option=com_bookmarks&task=detail&id=79.

CHAPTER

# 8

# Historical Influences on Fire Prevention

## OBJECTIVES

After reading this chapter you should be able to:

- Recognize significant fires that have occurred throughout U.S. history.
- Recognize the relationship between serious fires and resulting efforts to improve fire safety (and prevention) efforts.
- Recognize the significance of previous national planning efforts for the shape of fire prevention programs today.

Throughout history, fire has been a teacher for the human race. Fire has been a powerful force when harnessed for productive use, but its misuse or the inability to control it has led to disasters that have always provided lessons in how to become more safety conscious. Every fire that has gotten out of control could have taught someone something about what they could have done differently. But because most people do not believe they will ever have a fire, the lessons often come too late. And it is the larger, more **significant fires** that hurt many people that usually capture our collective attention.

significant fires
■ large fires with multiple deaths

Many large fires have resulted directly in changes to building and fire codes, including laws regulating construction or controlling behavior (such as smoking) that could cause fires. These laws are intended to keep fires from occurring, or to mitigate the damage if they do occur. Many prescribe changes in behavior in order to prevent fires from getting out of control.

This chapter briefly examines a few of the historic fires—predominantly in the United States—that have led to changes in codes and laws that reduce the risks from fires. The relationship between these fires and the changes they produced is simple, reinforcing the famous quote from George Santayana: "Those who cannot remember the past are condemned to repeat it."

# The Great London Fire

As reported in works available on the Luminarium Web site,[1] the Great London Fire began on September 2, 1666, in a bakery operated by the baker for King Charles II. Building construction methods of that time meant that most homes were built of wood and pitch—an excellent fuel load for the ensuing conflagration. Strong winds also helped spread the fire quickly.

The fire raced to the warehouse district where products such as hemp, oil, tallow, hay, timber, and coal contributed to the fuel load and the severity of the fire. The bucket brigades of the time could not keep up with the spread of the fire, which was only controlled in some areas by clear swaths created by a previous fire. By 8 o'clock that evening the fire had spread halfway across the London Bridge. As is still allowed by fire code today, the common way to stop large fires in those days was to destroy homes and buildings in the path of the fire—creating fire breaks. But by the time buildings were ordered to be demolished, the fire was already out of control. Eventually, buildings were ordered to be demolished with gunpowder, but the fuel load of demolished buildings could not be removed quickly enough to provide a fire break. The fire burned for three days when a final fire break finally aided in controlling the fire.

The Duke of York (later King James III) was credited with organizing and ordering firefighting efforts that helped control the fire, but some estimates counted 16 people dead but as much as 80 percent of the city buildings were destroyed.

This Great Fire and a subsequent fire in 1676, which destroyed another 600 homes south of the river, left a scar on London's collective consciousness. The only positive result was that the plague was greatly diminished because so many disease-bearing rats had been killed in the fire.

That blaze in old London, and subsequent fires there, taught the local decision makers the value of different construction features. Common sense, if nothing

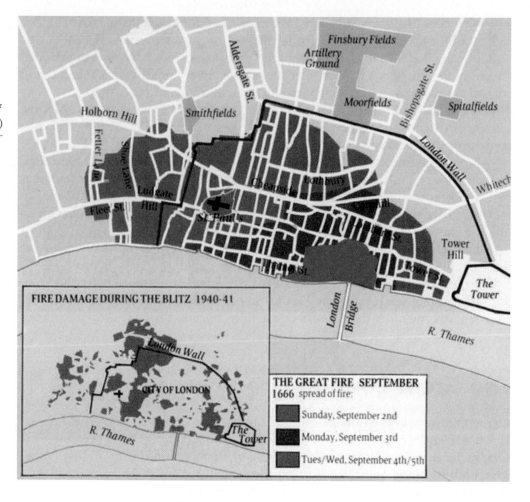

The Great Fire of London was an early example of how a significant fire could influence building codes and fire protection efforts. (*Courtesy of www.millwall-history.co.uk*)

else, would dictate that dwindling supplies of lumber and a history of fires should lead to more fire-resistive construction practices that could limit the quick spread of fires when they did occur.

## The Iroquois Theater Fire

In December 1903, a fire broke out in the 5-week-old Iroquois Theater in Chicago. Accounts reported in the *Chicago Tribune* helped portray the events as they unfolded.[2]

The bulding was called (at the time) "fire-proof" because of the fire resistive construction materials. However, the contents of the theater provided a great deal of fuel for the fire and proved that there is no such thing as "fire-proof" in a real sense. In other words, it is possible to build something that will withstand the ravages of fire—but impractical when the comforts that make up the interior are added.

Schools were out for Christmas break and some estimated at least 2,000 people (women and children) were inside the theater when the fire broke out. An arc light on the left side of the sage was listed as the likely cause, quickly igniting stage scenery. The actor and comedian performing at the theater (Eddy Foy) tried to calm the audience as fire started to spread. Stage hands tried to lower the curtain to control the flames, but it stalled. There were 27 exits provided for the theater but people trying to escape found many of them blocked and could not locate others. Later the investigation pointed out that an asbestos fire curtain was designed for stage fires, but had never been installed. According to the *Tribune* accounts, calls from fire officials asking if anyone was alive to make some kind of sound or groan went unanswered in the silence and heap of bodies left after the fire was controlled."[1]

The fire claimed more than 600 lives and injured another 250. The theater's managers and several public officials were indicted in connection with the fire, but in the end no one was actually punished.

The fire led to more stringent safety standards for theaters and other public buildings, including clearly marked exits and doors built so that even if they could not be pulled open from the outside, they could be pushed open from the inside. Today those exit-door features are commonly called "panic hardware."

# The Baltimore Conflagration

In February 1904, a fire started in a six-story mercantile building in downtown Baltimore, Maryland. According to reports on the GenDisasters Web site,[3] the Hurst & Co. dry goods building used a gasoline-powered engine as part of their operation, and though the fire began in the basement, the explosion from the gasoline engine is credited with causing a rapid fire that brought the entire building down quickly. The rapid spread of the fire quickly engulfed most of the wholesale district of Baltimore, destroying an entire commercial section of town including the Chamber of Commerce and the Stock Exchange. The fire was kept partially in control by a natural fire break (Jones Falls Creek) but moved further into the lumber district of the city and the docks.

Assistance in fighting the fire was provided by fire crews from New York, Philadelphia, Frederick, and many others—but the problem was that much of their equipment could not be used together. The firefighting hose threads were not standardized, and in many cases hoses could not be used. The fire eventually destroyed about 80 city blocks, some 2,500 buildings, and burned for 30 hours. Not a single life was reported lost in the fire, but the amount of damage led to rules that standardized the threads of firefighting hoses.

It is ironic that today, hose threads must be standardized, but fire hydrant and fire department hose connections do not have to be. According to representatives of the Knox Company, they still encounter situations where fire departments are not able to attach their hoses to fire hydrants or fire department connections (FDCs). This problem was identified because a few companies, such as Knox, have created locking caps for both hydrants and FDCs to decrease the risk that these devices will be taken off by vandals or those intent on selling the metal.

The Knox company can document instances in the United States recently where FDC caps were removed and garbage stuffed inside the pipes, and later attempts by fire departments to use them to supplement sprinkler system water flows from their fire engines were ineffective. In one case an entire apartment complex whose sprinkler system was out of commission was lost for those reasons.

The Baltimore conflagration of 1904 pointed out the need to standardize fire department hose threads. But recent efforts to protect fire department connections from vandalism have identified that there are still problems with standardization of thread patterns for fire department connections and fire hydrants. (*Courtesy of the Knox Company*)

# The Triangle Shirtwaist Fire

In March 1911, a fire broke out in the Asch Building in New York City. An excellent article posted on the Web site of the University of Missouri—Kansas City School of Law describes the events of the fire, and the legal battle that followed.[4]

According to author Doug Linder, the fire was relatively short-lived but intense. The Asch Building was home for the Triangle Shirtwaist Factory, which employed mostly teenage girls to make garments, many of them speaking little English. Though not a direct contributor to the fire losses, the language barriers were thought to have added to the confusion during the fire and after.

The start of the fire on the eighth floor of the building was attributed to some discarded rags between the fabric cutting tables. Attempts to throw pails of water on the fire did nothing to control it, and attempts by a shipping clerk to suppress the fire failed because the hose had no water pressure.

Personal accounts of the tragedy were chilling. Fire and dense smoke filled the fire floor and spread to the ninth and tenth floors, where one employee said he witnessed employees "running around like wildcats." Witnesses saw many young girls leaping from the tenth floor to their deaths below. Those who saw the same sight when the World Trade Center in New York was attacked will have a distinct visual image to compare the tragedy of that day in 1911.

Some employees were rescued via ladders left by painters to a nearby building. Some were able to escape by the stairs or a fire escape before it collapsed. Some escaped by leaping onto the elevator roof as it made its descent. But

The Triangle Shirtwaist fire of 1911 ultimately led to the establishment of one of the earliest building codes: NFPA 101, the Life Safety Code. (*International Ladies Garment Workers Union Archives, Kheel Center, Cornell University*)

145 employees, mostly young women, died in the fire or by leaping to their deaths to avoid it.

The fire damage was litigated for several years following the event, and eventually was settled in civil court—for about $75 per life, according to Mr. Linder. But this fire's ramifications for code changes are felt still today.

According to the fire marshal at the time, William Beers, the fire was most likely caused by a discarded match thrown either into waste near some oil cans or clippings under one of the cutting tables. There was a no-smoking policy throughout the company, but workers reported that smoking was commonplace. The building department was blamed, but they claimed that the building (with one fire escape) was built according to the codes at the time, and that they had insufficient funds to conduct regular building inspections.

Public pressure following the tragedy led to the formation of a commission on safety and working conditions in New York that produced a number of safety recommendations that included improved life safety measures in factories and loft buildings, better smoking control, fire drill requirements, and better exiting facilities for these types of structures. One member of the commission, Frances Perkins, was later appointed as the first Secretary of Labor under Franklin Roosevelt's administration and was a long-time advocate for workers' rights and safety. Ms. Perkins was the first female member of a presidential cabinet when she was appointed in 1933. In recalling the events of the Triangle Shirtwaist Factory disaster, Ms. Perkins later said that a strong sense of guilt motivated her and others to improve fire code regulations.

Ms. Perkins testified before the 1913 annual meeting of the National Fire Protection Association, imploring the association to adopt increased safety standards. It was largely as a result of her efforts that the NFPA formed the Safety to

Life Committee and published the first version of NFPA 101—the Life Safety Code—which was originally known as the buildings exit code.

# The Our Lady of the Angels School Fire

Ninety students and three Catholic nuns died in the fire at Our Lady of the Angels School in Chicago in December 1958. Another 77 were injured as the fire spread throughout the hallways and stairwells of the school, blocking escape paths. A report from the NFPA Quarterly of January 1959 (Chester I. Babcock and Rexford Wilson authors) examined the major facts of the fire.

The school was older, having been built in 1910. And as is still common today, it was in compliance with codes at the time it was built, but lacked some of the safety features required by the codes of the day. There were no fire sprinklers or automatic fire alarm systems. There was only one fire escape, and reaching it required some building occupants to move through the main corridor—which was filled with smoke by the time the fire was under way.

The exterior of the building was brick, but the interior was constructed of wood, including the exit paths and the roof. The floors were wood and had been coated many times with flammable petroleum-based waxes, which investigators felt contributed to the fire load and its quick spread.

The fire began in the afternoon while children were still in class, but when a teacher and two of her students smelled smoke, they tried to consult with the principal before beginning an evacuation. By the time the students and teachers began exiting the building and the fire department was notified, some estimated that the fire had burned for anywhere from 15 to 30 minutes. After the teacher evacuated her students, she went back inside the school and operated the fire signal.

First-floor students and teachers used five available stairways to exit the building. Second-floor occupants were cut off by the smoke that came up through an open door in the fire separation wall. Increasing heat broke out a window at the foot of a stairwell, providing an influx of fresh oxygen and fueling the fire even more. Heavy fire doors protected the first floor, but were not in place for the second floor, contributing to the spread of smoke and heat.

The cause of the fire was not officially determined, but sources suspected that a child smoking ignited combustibles stacked under a stairway. In all, 329 children and 5 nuns were left to escape the second floor by jumping from the windows to concrete or crushed rock below.

Though there had been many school fires before the Our Lady of the Angels, that fire is generally credited with leading to more stringent fire and building code requirements because of the multiple student deaths. And for those who have often wondered why schools usually receive more fire inspections than other properties, the fire at Our Lady of the Angels provides much of the rationale.

By comparison, another school fire in Kenilworth, Illinois, in 1958 provided much different results. That fire was handled very differently by staff, who activated a more sophisticated alarm system that led to a quick evacuation of the building and notification of the fire department. Fire sprinklers in a closet were credited with controlling the fire.

# The MGM Grand Hotel Fire
# (and Las Vegas Hilton)

On November 21, 1980, the MGM Grand Hotel in Clark County, Nevada, experienced a serious fire that resulted in 85 fatalities and injured about 650 more. The hotel, opened in 1973, had about 2 million square feet of space. The ground-floor casino and showroom was very large—bigger than a football field—and was not required by code to have fire sprinklers, though other portions of the building did.

The fire began in a restaurant in an unsprinklered area on the casino level. According to an official investigation report, the fire spread within 6 minutes to the entire casino area. The fire originated in a concealed space in the restaurant because of an electrical problem that had evidently been brewing for a long time. Uninsulated wires for a refrigeration unit were being stretched and rubbed over a long period of time by the vibration of the appliance.

Once the fire got started, it moved quickly into the ceiling and air return system above the casino. Once there, it was fueled by flammable furnishings, including wall coverings, PVC plastic piping, glue, and plastic mirrors on the walls. Smoke traveled throughout the hotel, trapping many of the occupants on higher floors, where escape was difficult if not impossible. Fire officials found bodies of those who tried to escape in the inoperable elevators where they had tried to claw their way out but were overcome with smoke and soot. Fourteen firefighters were also hospitalized from the fire, mostly from smoke inhalation.

A 2005 account published by the *Las Vegas Review-Journal,* written by Jane Anne Morrison, recounted some of the lessons learned from the fire. Ms. Morrison wrote that despite pressure from fire marshals during the hotel's construction in 1973, hotel executives fought the installation of fire sprinklers. Sprinklers would have added about $192,000 to the cost of the $106 million hotel. "With sprinklers, it would have been a one- or two-sprinkler fire and we never would have heard about it," said David Demers, the Massachusetts fire analysis specialist who co-authored the NFPA report on the fire.

But a lack of fire sprinklers was not the only problem. The U.S. district judge involved in the extensive litigation after the fire was known to carry a keychain with a fusible link from the smoke control system that was allegedly installed incorrectly and allowed the smoke to travel throughout the hotel. The fire alarms were manual and did not operate because no one pulled them. Some speculated that might have been a blessing in disguise, because had they operated, they might have caused even more people to rush into smoke-filled hallways.

The inevitable litigation dragged on for years after the fire and has been largely credited with increasing pressure on hotel operators to retrofit fire sprinkler systems and improve their safety measures. There were some 1,327 lawsuits filed against 118 companies. Money from many of them went into a $223 million settlement fund that was distributed to victims and their families. The $105 million it cost the MGM Grand was the largest, followed by $14.4 million paid by Simpson Timber Company for providing below-grade ceiling tiles and flammable adhesive. But that process took years, and in the short time following the MGM Grand fire, opposition to retroactively installing fire sprinklers continued. The retrofit costs were estimated at about $2 million for a typical Las Vegas Strip hotel.

But opposition dissipated when an arsonist set fire to the Las Vegas Hilton Hotel on February 10, 1981. Eight people were killed and another 200 injured in that fire. The timing of that fire coincided with a national press presence in Las Vegas covering the fact that Frank Sinatra was going before the gaming commission to obtain a gaming license. News of that fire spread very quickly, and opponents of retrofitting fire sprinklers were marginalized by heavy public opinion in favor of fire sprinklers.

The effort to increase safety for hotels was significantly affected by these fires, but resistance to retrofitting other buildings with fire sprinklers remained.

## The Bhopal, India, Chemical Disaster

Fire safety regulations do not always arise out of fire tragedies. The chemical disaster that occurred in Bhopal, India, in December 1984 created significant pressure in the United States. The incident occurred at a Union Carbide plant in Bhopal. According to accounts posted on sites such as wired.com,[5] the incident began with a chemical reaction. Methyl isocyanate (MIC) gas was released when water leaked into one of the storage tanks on the night of December 2. Gas began leaking but an alarm evidently did not activate for 2 hours. Those nearest the plant felt the effects of the blinding noxious gases that damaged their lungs and induced vomiting.

Exact numbers of those affected has never been agreed upon. But in a community of 900,000—some workers claimed to have cleared 15,000 bodies of those who perished in one of the worst industrial accidents in history. The effects of huge disasters like oil leaks have lasting environmental and economic damages. But few if any have ever matched Bhopal in terms of deaths. And some estimate than another 50,00 at a minimum have become invalids or have developed serious respiratory conditions as a result of the leak. When political leaders realized that similar events could occur in the United States, they pushed for legislation designed to prevent such an event. The Superfund Amendments and Reauthorization Act (SARA), Title III, created a legal mechanism that required businesses using or storing hazardous materials to notify the community, and in particular emergency responders. This unfunded mandate created a heavy burden on the U.S. Fire Service. As more and more hazardous materials operations were identified, the workload imposed by keeping up with the amount and type of hazards they presented mounted. Many local departments resorted to charging fees to support the inspection services required of the communication methods (including databases listing hazardous material types, quantities, and locations) to inform emergency responders of the hazards they might face should a fire occur in one of these businesses.

## The First Interstate Bank Fire

In May 1988, fire broke out at the First Interstate Bank tower in downtown Los Angeles. During the late evening of May 4 and the early morning of May 5, members of the Los Angeles City Fire Department successfully stopped one of the most devastating high-rise fires in the history of Los Angeles or the United States. According to a report produced by the U.S. Fire Administration[6]:

> Extinguishing this blaze at the 62-story First Interstate Bank Building required the combined efforts of 64 fire companies, 10 city rescue ambulances, 17 private ambulances, 4 helicopters, 53 command officers and support personnel, a complement of 383 firefighters and paramedics, and considerable assistance from other city departments. The fire destroyed four floors and damaged a fifth floor of the building, claimed one life, injured approximately 35 occupants and 14 fire personnel, and resulted in a property loss of more than $50 million.
>
> There were many factors that contributed to the fire and its spread.

- Building security and maintenance personnel delayed notifying the fire department for 15 minutes after first evidence of the fire.
- Smoke detectors on several floors had been activated, and deactivated several times before notifying the fire department.
- Fire sprinklers were installed in 90 percent of the building, but operating valves had been closed awaiting installation of water flow alarms.
- The fire originated in an open-plan office area in the southeast quadrant of the 12th floor. The area of origin contained modular office furniture with numerous personal computers and terminals used by securities trading personnel. The cause is thought to be electrical in origin, but the precise source of ignition was not determined. The fire extended to the entire open area and several office enclosures to fully involve the 12th floor, except for the passenger elevator lobby, which was protected by automatic closing fire doors.
- The fire extended to floors above, primarily via the outer walls of the building; windows broke and flames penetrated behind the spandrel panels around the ends of the floor slabs. The curtain wall construction creates separations between the end of the floor slab and the exterior curtain wall.
- There was heavy exposure of flames to the windows on successive floors as the fire extended upward from the 12th to 16th floors. The flames were estimated to be lapping 30 feet up the face of the building. The curtain walls, including windows, spandrel panels, and mullions, were almost completely destroyed by the fire. There were no "eyebrows" to stop the exterior vertical spread, and fireground commanders were concerned about the possibility of the fire "lapping" higher to involve additional floors.

Fire companies in Los Angeles were credited with making a tremendous effort to stop the spread of the fire—and doing so in an environment where the fire was coming up toward them. There were essentially no fire sprinklers at the time of the fire.

Though this fire is not viewed as important (or mentioned often) compared to the MGM Grand Hotel Fire, the First Interstate Bank Tower fire in Los Angeles was used by fire departments in many jurisdictions across the nation as an example of the need for fire sprinklers. Many fire departments (including Louisville, Kentucky) conducted studies of their own to demonstrate that even many larger jurisdictions in the nation lacked sufficient firefighting resources to control a high-rise fire. Louisville actually demonstrated that parking a full complement of firefighters and equipment at the base of a high-rise building—eliminating the problems of response time—would not allow them to get to a high floor in time to prevent flashover from occurring.

Because many fire departments could not expect to successfully control a high-rise fire, there were numerous retrofitting ordinances for high-rise buildings across the nation to require fire sprinkler systems. That concept was driven home by another fire that occurred some years later.

# The One Meridian Plaza Fire

The One Meridian Plaza Fire in Philadelphia occurred on February 23, 1991.[7] The fire received more national attention than the First Interstate Bank fire because three firefighters were killed battling the blaze that began on the 22nd floor of the 38-story Meridian Bank Building. In all there were 51 engine companies, 15 ladder companies, 11 specialized units, and more than 300 firefighters on scene. The fire destroyed eight floors of the building, and notably it was finally controlled only when it reached a floor that was protected with automatic fire sprinklers.

The estimated $100 million in direct property loss was just the tip of the iceberg. The indirect losses—including 15 seriously injured firefighters, lost business at the plaza itself, and lost revenue to the surrounding businesses that relied on the high-rise occupants for their own business—counted into the hundreds of millions of dollars.

According to the article, litigation resulting from the fire ended up producing an estimated $4 billion in civil damage claims.[4]

That fire not only destroyed the building, it forced many of the smaller businesses nearby to close their doors. They depended on commerce in that area to survive, and when the fire at One Meridian Plaza shut down their primary source of business, they could not survive. National estimates are not reliable, but anecdotal evidence leads those in the fire prevention profession to recognize that after a serious fire, a very high percentage of businesses do not reopen. This was the first case that called attention to a serious fire actually causing others to go out of business as well.

Both the First Interstate Bank Tower and the One Meridian Plaza Fires highlighted the limitations of most cities in fighting a high-rise fire and created a movement across the United States to retrofit fire sprinkler systems in high-rise buildings.

# The Oakland Hills Fire

On October 20, 1991, a fire began in the hills outside of Oakland, California. Investigators determined the cause as embers from a previous fire in the same location, which were fanned by wind and traveled outside the perimeter the fire companies had established. According to an account published online by Captain Donald R. Parker,[8] the fire began despite heavy overhaul by the companies on scene before the fire began. They had also placed hose lines along the perimeter and had checked on the fire scene several times during the night before.

The ensuing blaze grew quickly, and overwhelmed the resources on hand. Numerous problems hampered firefighting—including, nonstandard hydrant connections that made it difficult for supporting fire departments to assist. The roads leading into the area were quickly clogged by people trying to escape and fire crews trying to get in to the fire scene. Some curious observers evidently made that situation even worse.

Over the next 72 hours the fire raged before being brought under control. In the end, 25 people were killed and another 150 were injured. Nearly 3,500 homes were destroyed, the fire had consumed more than 1,500 acres, and the damage estimates were listed as $1.5 billion.

There have been other serious urban or wildland interface fires before and since, but this fire pointed out several problems and helped galvanize national efforts to create safety strategies—including defensible space around homes and noncombustible roofing materials—designed to minimize damage during these types of fires.

Currently, NFPA manages a program for urban wildfire interface called "Firewise." The International Code Council also is producing model codes and extensive work in this arena, and the U.S. Fire Administration is also promoting safety efforts including a program titled "Ready, Set, Go" in partnership with the International Association of Fire Chiefs and NFPA.

# 9/11

On September 11, 2001, the United States experienced the most aggressive foreign attack on U.S. soil since the bombing of Pearl Harbor. Another site, the Pentagon, was also attacked, and it survived some serious damage though many were killed. The main focus of the national tragedy was the World Trade Center in New York City, two of the tallest buildings in the United States. Damage was not limited to the Twin Towers. Hijackers deliberately flew two commercial passenger jets into the towers, setting off fires in each and in some of the surrounding buildings.

The terrorist attack on September 11, 2001, proved to be one of the worst disasters to befall the United States. Even though its cause, terrorism, was extremely rare, the event led to increased focus on fire protection of steel structural supports and exiting of high rises. (*Mike Rieger/ FEMA News Photo*)

Nearly 3,000 people, mostly civilian, lost their lives in the fires and ensuing building collapse on that day. That number includes more than 300 firefighters who gave their lives attempting to evacuate and rescue those still trapped in the buildings.

People from around the world watched in horror as the buildings burned—and the gradual realization sunk in that they were going to collapse. Witnesses saw people jumping from upper floors, just as they had in the Triangle Shirtwaist fire in 1911, but in far larger numbers. People who jumped surely realized they would die, but preferred to go that way rather than burning alive in a fire. When the buildings did collapse, many more people died, and many who had escaped were further injured by the falling debris. At least one death was attributed to lung problems from the concrete dust long after the event. The survivors and the family members of those lost in the tragedy still suffer from the physical and emotional damage caused by that event, and the obvious malice behind it.

Possible changes in building codes related to the 9/11 tragedy are still being debated today. The most comprehensive study on the incident related to building and fire codes was conducted by the National Institute of Standards and Technology (NIST). It is an oversimplification of the issues involved, but many were concerned that a high-rise building could be caused to collapse under fire conditions, and there was some call to strengthen building codes to prevent such an event.

The building construction of the World Trade Center towers used protected steel for structural support. These steel support elements were supposed to withstand the effects of fire in a building—but no one anticipated that a jet airliner flying into a building would shear off the protective covering on the steel while also subjecting the building to a far greater fuel load (from the jet fuel) than it would normally experience from fire. The resulting collapse caused NIST, the scientific engineering community, and building and fire officials to rethink the way they saw protection of steel elements, and how much exiting to provide in a high-rise building.

A full report can be viewed on the NIST Web site that deals more comprehensively with the issues surrounding this tragedy.

# The Station Nightclub Fire

On February 20, 2003, fire broke out at the Station Nightclub in West Warwick, Rhode Island. The fire occurred as a result of indoor pyrotechnics being used as part of a rock band's performance at the club that night. The fire began when the pyrotechnics ignited some flammable coverings on the ceiling and quickly spread throughout the building.

In that fire, 100 people died, some 230 were injured, and 132 escaped uninjured. There were a number of factors that contributed to the fire and its spread.

First, and perhaps most obvious, the indoor pyrotechnics were set off illegally, without the benefit of an inspection or a permit. The flammable coverings on the walls and ceilings were illegally installed to help with the acoustics of the venue and improve the sound of the frequent performances. The club was a popular place for dancing and music, attracting hundreds who were there to eat, drink, and party.

The Station Nightclub fire in Rhode Island in 2003 provided a number of lessons about flammable coverings in public occupancies, the need for fire sprinklers, the need for frequent qualified inspections, and the effect of human behavior on fire exiting. (*Courtesy of NIST*)

Previous fire code inspections had missed the fact that the coverings were illegal and would contribute to rapid flame spread. The building was not required to have fire sprinklers, and it did not have them. It did comply with existing codes for exiting, but a NIST report on the topic estimated that about two thirds of the people who were trying to escape the fire did so by trying to go out the front door—the way they had come in.[9]

According to the NIST report, measurements produced in a fire test conducted on a mockup of a portion of The Station nightclub platform and dance floor produced—within 90 seconds—temperatures, heat fluxes, and combustion gases well in excess of accepted survivability limits.

A computer simulation of the full nightclub fire suggests that conditions around the dance floor, sunroom, and dart room would have led to severe incapacitation or death within about 90 seconds after ignition of the foam for anyone remaining standing in those areas—and not much longer even for those close to the nightclub floor.

In other words, because of the fast-moving fire, people had little time to escape.

Legal battles inevitably followed the event. The State Fire Marshal of Rhode Island and the local fire chief narrowly escaped criminal prosecution for their alleged failure to provide for adequate safety inspections of the venue. The civil suits still move forward today, and, as usual, any number of parties are named. The club manager was convicted by a court of criminal negligence and sentenced to a prison term of 4 years. Following his trial, the Station's owners were scheduled to receive separate trials. However, on September 21, 2006, Superior Court

Judge Francis J. Darigan announced that the brothers had changed their pleas from "not guilty" to "no contest," thereby avoiding a trial.[5] Michael Derderian received 15 years in prison, with 4 to serve and 11 years suspended, plus 3 years probation—the same sentence as the band manager. Jeffrey Derderian received a 10-year suspended sentence, 3 years probation, and 500 hours of community service.

The tragedy of that fire led to retroactive sprinkler requirements for public assembly occupancies all across the nation, and changes to the NFPA 101 Life Safety Code. Those changes made fire sprinkler requirements more stringent for nightclubs, basing their requirement on fewer occupants than previously. The lessons learned have been a teaching tool in fire prevention training programs throughout the United States, greatly increasing fire departments' awareness of the need to make sure their inspectors are properly trained and that inspection cycles are as short as possible.

# Historical Planning Efforts

**historical planning efforts**
■ national planning processes that have led to improvements in fire prevention in the United States throughout our history

**Historical planning efforts** refers to landmark plans developed in part because of the lessons learned from historical disasters. These national strategic plans have attempted to shape the direction of fire protection and prevention efforts throughout U.S. history.

If history's disasters have taught us something about how to improve fire prevention efforts, then the resulting planning efforts have as well. Chapter 7 deals with planning for fire prevention programs, but what of national planning processes? What significance do they have for fire prevention efforts today?

The first national planning process for fire prevention efforts for the United States occurred on October 13–18, 1913, in Philadelphia. It was the first documented effort to collect information on national fire loss data and to begin doing something about it. Far more sophisticated, and noteworthy, was President Truman's conference on fire prevention in 1947. The U.S. Fire Administration maintains a Web site that includes the report produced by this conference. In the press release announcing the conference, we find the historical links that brought this conference about:

> For more than a decade the loss of property in the United States due to fires has been steadily mounting year by year. During this period an average of 10,000 persons have been burned to death or have died of burns annually. In the first nine months of this year fire losses reached the total of nearly half a billion dollars, with the prospect that final reports for 1946 will show this year to have been the most disastrous in our history with respect to fire losses.
>
> Additional millions must be added to the nation's bill because of forest fires which, in 1945, accounted for the destruction of more than 26 million dollars worth of timber, a precious national resource. Also must be added the enormous sums spent in fighting and controlling fires.
>
> This terrible destruction of lives and property could have been almost entirely averted if proper precautions had been taken in time. Destructive fires are due to carelessness or to ignorance of the proper methods of prevention. These techniques have been tested, but they must be much more intensively applied in every State and local community in the country.

The President has, therefore, decide to call a National Conference on Fire Prevention, to be held in Washington within the next few months, to bring the ever-present danger from fire home to all our people, and to devise additional methods to intensify the work of fire prevention in every town and city in the Nation.

He has appointed Major General Philip B. Fleming, Administrator of the Federal Works Agency and of the Office of Temporary Controls, to serve as general chairman of the conference. General Fleming, who served in a similar capacity during the President's Conference on Highway Safety last May, already is at work on preliminary arrangements for the meeting, to which will be invited State and local officials who have legal responsibilities in the matter of fire prevention and control, and representatives of non-official organizations working in this field.

The new impetus given to the prevention of traffic fatalities by the Highway Safety Conference already has resulted in saving several thousand lives, and the benefits will continue to be felt as the techniques adopted by the conference are increasingly applied. The President is encouraged to hope, therefore, that a similar attack on fire losses will yield corresponding benefits.

Indeed, that the taking of proper precautions can stem this staggering drain on our resources is well illustrated in our experience with the Nation's forests. Although the acreage of our unprotected forest lands amounts to only 25 percent of the acreage of our protected forests, the losses of the former in 1945 exceeded those of the protected tracts by more than 20 percent.

The President said: "I can think of no more fitting memorial to those who died needlessly this year in the LaSalle Hotel fire in Chicago, the appalling disaster at the Winecoff Hotel in Atlanta, and the more recent New York tenement holocaust than that we should dedicate ourselves anew to ceaseless war upon the fire menace."

The impetus for the conference came from fire disasters that occurred in the United States in a relatively close time frame to the conference. It was a presidential decree that the conference take place, which was a reflection of its historic proportions. But the fact that the president was motivated in part by recent disasters and previous successes from another safety initiative provides us with some perspective on how historic events shape public pressure to focus on fire prevention.

Another landmark report came from a commission appointed by President Richard Nixon in 1972. The National Commission on Fire Prevention and Control, led by Richard Bland, from Pennsylvania State University, was titled "America Burning" and was published in 1973. That report stated:

> The striking aspect of the Nation's fire problem is the indifference with which Americans confront the subject. Destructive fire takes a huge toll in lives, injuries, and property losses, yet there is no need to accept those losses with resignation. There are many measures—often very simple precautions—that can be taken to reduce those losses significantly. The Commission worked in a field where statistics are meager, but its estimates of fire's annual toll are reliable: 12,000 American lives, and more than $11 billion in wasted resources. Annual costs of fire rank between crime and product safety in magnitude. These statistics are impressive in their size, though perhaps not scary enough to jar the average American from his confidence that "It will never happen to me." In a Washington hearing the Commission heard testimony from the parents of a . . . boy who caught fire after playing with matches. They described the horror of the accident, the anxiety while awaiting doctors' reports, the long weeks of separation during the critical phases of treatment, the child's agony during painful treatment, the remaining scars, and the many operations that lie ahead. Multiply that experience by the 300,000 Americans who are injured by fire every year, and consider,

as we did, that it could easily happen in your own family; then the Nation's fire problem becomes very immediate and very fearsome. During its deliberations the Commission uncovered many aspects of the Nation's fire problem that have not received enough attention—often through indifference, often through lack of resources. It became clear that a deeper Federal involvement was needed to help repair the omissions and help overcome the indifference of Americans to fire safety.

That report galvanized fire protection interests in the nation, motivated once again by the losses the nation experienced from fire. In this case no specific incidents were identified; rather, an accumulation of fire losses indicated the nation was losing too many lives, and too many of its resources, to fire.

The Commission was staffed by a man named Howard Tipton, who became the first administrator of the first federal agency to be formed specifically for fire prevention and control. The National Fire Prevention and Control Administration was established in 1974 and later became the U.S. Fire Administration, which included the establishment of the National Fire Academy. The U.S. Fire Administration maintains a copy of the report "America Burning" on its Web site at www.usfa.dhs.gov.

Since that time, numerous national conferences have been conducted that have attempted to galvanize the nation's fire service—if not the general public—toward fire prevention and control. The Wingspread conferences, named after the Wisconsin location where they were conducted occurred during the 1960s, '70s, and '80s, preceded and followed "America Burning." These conferences brought together national fire service leaders to wrestle with issues facing the fire service, and inevitably ended up recommending increased emphasis on fire prevention efforts as a way to improve public safety in the United States.

More recently, a grant from the Department of Homeland Security allowed the Institution of Fire Engineers to conduct another national fire prevention planning conference, "Vision 20/20." In March 2008, 170 national fire service leaders and others from outside the fire service came together to discuss what should be done to carry on previous efforts to reduce fire losses in the United States. That conference, and the accompanying report, is yet unproven in its impact on the nation's fire problem.

The conference envisioned five separate strategies to further improve fire prevention efforts in the nation—many of them repeats from previous reports. A copy of the report may be found on the Web site for the planning effort and time will tell whether the strategies advocated by the conference actually gain enough support to produce results. But Vision 20/20 is just the latest example of how the national fire services, and others devoted to fire safety, come together to try and move fire prevention efforts forward collaboratively.

A compilation of all of the recommendations in these reports was conducted by Woody Stratton of the U.S. Fire Administration, and Gary Keith of the National Fire Protection Association, as part of the preplanning work done for Vision 20/20. Progress has indeed been made in the area of fire prevention and control. In the 1970s about 12,000 people were dying each year from fire and burns in the United States. Today, about 3,500 are dying annually, according to figures published by the NFPA and the U.S. Fire Administration. But fire losses in the United States continue to be higher than in other industrialized nations, and significant fires still occur that help to galvanize public will to improve our nation's efforts to prevent fires and to control our losses when they do occur.

# Summary

Fire tragedies and other disasters have led in some cases to the establishment of improvements in building and fire codes, and in others to changes in technology and behaviors that have made the nation more fire safe. These historical influences have had a profound effect on where we are today with regard to fire prevention efforts. A partial list of significant fires, some of them not mentioned in this chapter, and their accompanying impact on fire prevention efforts would include:

- The Iroquois Theater fire of 1903—leading to code changes for exiting, fuel reduction, and panic hardware assemblies for exits in public occupancies
- The Baltimore conflagration of 1904—leading to the creation of standardized fire department hose threads
- The Triangle Shirtwaist fire of 1911—leading ultimately to the establishment of the NFPA Safety to Life Committee and the publication of NFPA 101, the Life Safety Code
- The Cocoanut Grove fire of Boston in 1942—leading to code changes for emergency lighting and general awareness of public assembly fire safety
- The Winecoff Hotel Fire in Atlanta, Georgia, in 1946—which led to improvements in fire protection features in Georgia, but was also cited as a motivation for the president's national conference on fire prevention in 1947
- The Texas City explosion of 1947—highlighting the dangers of ammonium nitrate, and leading to rules regulating the transportation of hazardous materials

- Our Lady of the Angels school fire of 1958—leading to more aggressive inspection practices in schools
- The Beverly Hills Supper Club fire in Southgate, Kentucky, in 1977—leading to code enforcement responsibilities across state lines
- The Station Nightclub fire in Rhode Island in 2003—which led to greater emphasis and increased fire sprinkler requirements for smaller public assemblies, especially nightclubs

A comprehensive list of significant events in the history of fire protection in the United States, called "Key Dates in Fire History," is maintained on the NFPA Web site. Each event has lessons to teach us about how fire safety efforts may be improved.

Previous national planning efforts for fire prevention are also part of the historical influence on fire prevention efforts. Every planning effort since the first one recorded, in 1913, has influenced the direction of fire prevention efforts in the United States.

For people involved in fire prevention efforts, an understanding of how those efforts have evolved will provide lessons about how to improve them. Many of the recommendations that came from significant fires, other events, and national prevention planning efforts have striking similarities. This may tend to discourage those who feel we are just repeating the same old things, ultimately fulfilling the axiom that the definition of insanity is repeating the same action over and over but expecting different results. However, prevention efforts in the United States have improved, and may continue to do so with continued emphasis at the national, state, and local level.

# Case Study

The City of Taylor, Maryland, has maintained a fire prevention program since it was formed in 1913. The fire marshal of Taylor (Chief Stevens) in 1914 established an aggressive home fire safety survey, which was credited with controlling fire loss rates in that community and was supported heavily for years. However, when that fire marshal retired in 1918, the program died, and emphasis on fire prevention diminished for Taylor.

Over the years, the emphasis on fire prevention programs rose and fell, but continued a generally upward trend, and in the 1970s, a comprehensive approach to fire prevention was established in Taylor. Prevention efforts included active participation by the fire department in plan review for new construction—focusing on fire exiting, fire sprinkler and alarm systems, fire department access, and water supply for firefighting purposes. They also included an aggressive fire code enforcement program, with sufficient resources to conduct annual code compliance inspections of every business occupancy in the city.

Concurrently, the City of Taylor has maintained full-time staff for fire investigation efforts and developed a close working relationship with the police department and the local prosecutor's office to improve their ability to identify and prosecute the crime of arson. Finally, Taylor has been nationally recognized for its public education efforts, receiving the first Award of Excellence from the International Association of Fire Chiefs in 1989, for an aggressive door-to-door public education campaign in a largely African American portion of that city that included an emphasis on working smoke alarms.

Since 1988, the fire prevention programs in Taylor have been faced with budget reductions. Local tax "revolts" have led to the establishment of property tax limits, and a gradual reduction of revenue for city services. Because the fire prevention programs are funded out of the same pool of money that pays for emergency operations, a typical budget cycle finds prevention programs pitted against emergency operations for limited funding that is designated for the Patterson fire department.

Today, the fire marshal of Taylor is being faced with budget cuts and has been asked to justify all of the department's prevention efforts, but especially the code enforcement efforts in existing business occupancies. Local business leaders have provided copies of statistical data from the annual fire department report that indicates most fires, fire-related injuries, and deaths occur in residential occupancies. Most of those occur in one- and two-family dwellings that are not regulated by the fire code.

Faced with active cuts in regular business inspection resources, the fire marshal has been challenged to identify the rationale for maintaining regular business inspection practices.

# Case Study Review Questions

1. How could an understanding of the historical influences on fire prevention efforts be used to aid the fire marshal's defense of code compliance inspections?
2. Describe the relationship between significant fires and the need for regular code compliance inspections.
3. Given the case study, explain how significant fires might be used to support efforts in any local jurisdiction.
4. Explain how one might identify the relationship between local data and anecdotal examples of significant fires to advocate for aggressive code compliance efforts.
5. Explain how one might use the history of significant fires in the United States to promote the involvement of the Patterson fire department in the promulgation of national fire codes and standards.
6. Explain how previous national planning efforts have helped shape modern fire prevention efforts.
7. What problems did the Iroquois Theater fire point out?
8. What problems did the great Baltimore Conflagration point out that may still exist today?
9. What problems did the Station nightclub fire point out?
10. What did it take in addition to the MGM Grand Fire in Las Vegas to tip political support in favor of retrofitting fire sprinklers in high-rise buildings?

# Endnotes

1. N.A. *The Great Fire of London, 1666*. Retrieved from http://www.luminarium.org/encyclopedia/greatfire.htm.

2. www.chicagotribune.com—reporter Bob Secter.

3. Beitler, S. (posted May 27, 2008). Baltimore, MD. *The Great Fire Feb 1904*. Retrieved from http://www.gendisasters.com/data1/md/fire/baltimore-firefeb1904.htm.

4. Linder, D. (posted 2002). *The Triangle Shirtwaist Fire Trial*. Retrieved from http://www.law.umkc.edu/faculty/projects/ftrials/triangle/trianglefire.html.

5. Long, T. (posted December 3, 2008). Dec 3, 1984: *Bhopal, Worst Industrial Accident in History*. Retrieved from http://www.wired.com/science/discoveries/news/2008/12/dayintech_1203.

6. N.A. (May 1998). U.S. Fire Administration/Technical Report Series: Interstate Bank Building Fire. Los Angeles, CA. USFA-TR-022. Retrieved from http://www.usfa.dhs.gov/downloads/pdf/publications/tr-022.pdf.

7. N.A. (n.d.). *High-rise Office Building Fire, One Meridian Plaza, Philadelphia, Pennsylvania*. Retrieved from http://www.iklimnet.com/hotelfires/meridienplaza.html.

8. Parker, D. (January 1992). *The Oakland-Berkeley Hills Fire: An Overview*. [Oakland Office of Fire Services.] Retrieved from http://www.sfmuseum.org/oakfire/overview.html.

9. N.A. (n.d.). *The Station Nightclub Fire*. Retrieved from http://en.wikipedia.org/wiki/The_Station_nightclub_ fire.

# 9
# Social and Cultural Influences on Fire Prevention

Culture *p. 169*
Cultural anthropology *p. 169*

Diversity and fire prevention *p. 180*
Generational culture *p. 180*

Multicultural society *p. 169*

## OBJECTIVES

After reading this chapter you should be able to:

- Identify U.S. cultural beliefs and traditions that influence fire prevention.
- Identify relationships between cultural beliefs and traditions in other countries and how they affect prevention.
- Describe how the United States is becoming a multicultural society.
- Describe the fire service culture and its impact on fire prevention.
- Describe how multicultural beliefs and traditions may be an advantage for fire prevention efforts.

PEARSON

For additional review and practice tests, visit **www.bradybooks.com** and click on MyBradyKit to access book-specific resources for this text!

It should be no surprise to anyone that the United States is becoming a **multicultural society**. More will be said about that topic later in the chapter—but before we can understand that, we must understand the term **culture** and how it affects prevention efforts.

*Culture* refers to the beliefs, norms of behavior, and traditions that form the social glue of a people. Those beliefs and traditions may be very different, depending on the culture. For example, East Indian culture is very different from Irish—in the food that is eaten, in their dress, in their approach to dealing with love and marriage. A multicultural society, then, is one that is made up of numerous cultures, all maintaining some sense of cultural identity while melding certain aspects of their cultures into a common set of traditions and beliefs. For example, there are many cultures in the United States, but what they have in common is a belief in personal freedom and a democratic society where people have power over government.

Dictionary.com defines culture in part as "a particular form or stage of civilization, as that of a certain nation or period" (e.g., Greek culture); or "the behaviors and beliefs characteristic of a particular social, ethnic or age group" (e.g., the youth culture, the drug culture, Asian culture). When we talk about culture, we are really talking about the beliefs, the values, the traditions, and the behaviors that make up the norm for a specific subset of any given society.

Sometimes culture refers to the society as a whole (as in Greek culture), and other times to subsets of an overall society (as in the youth culture or organizational culture). In every case, however, we are talking about unwritten rules that bind people into a relative identity. The concept of culture is a topic of its own, and **cultural anthropology** is a field of study that goes well beyond fire prevention. For readers who are interested in a more detailed discussion, the Society for Cultural Anthropology produces a valuable resource, an online journal that provides in-depth information on the topic of cultural anthropology. Anthropological studies provide us with insight into how a particular group of people might respond to a variety of situations. Thus, there are aspects of culture and anthropology that have a direct bearing on fire prevention. This chapter provides a brief overview of cultural factors relevant to the field of fire prevention, including how culture may increase or decrease fire risk; how history shapes a particular culture; how the United States is becoming an increasingly multicultural society; how the fire service culture affects prevention efforts; and how some solutions are emerging that demonstrate the value of a multicultural society for prevention efforts.

**multicultural society**
■ one that is made up of numerous cultures—all maintaining some sense of cultural identity—while melding certain aspects of their cultures into a common set of traditions and beliefs

**culture**
■ the beliefs, norms of behavior, and traditions that form the social glue of a people

**cultural anthropology**
■ the science that deals with the origins, physical and cultural development, biological characteristics, and social customs and beliefs of humankind

# Cultural (Social) Impacts on Fire Risk

Culture affects fire risk through traditions, customs, and beliefs. These are learned through experience and passed on from generation to generation. They govern the way people behave, and that has a direct impact on fire risk. What people believe about fire is an essential factor in how they handle it. For example, some Native American people view fire in a fatalistic sense—any attempt to deal with it proactively (as in prevention) will only ensure that an out-of-control fire will in fact occur. How these Native Americans view fire is a cultural belief that affects their ability (and ours) to deal with prevention, and can increase the risk of fire.

## CULTURE AND INCREASED FIRE RISK

How people use fire is also a factor in fire risk. Those who cook with open fires may be at greater risk than those who do not. In the late 1970s, after the Vietnam War, many Asian immigrants entered the United States. Many had grown up using small open fires to cook and knew of no other way to do so. Coming to the United States and discovering electric or gas ranges and ovens—let alone microwaves—was a cultural shock they were (in many cases) not prepared to manage.

Many of these immigrants continued to cook with open fires. A commonly available apparatus to do so was a small charcoal cooking grill. The problem many of them encountered was using such a grill indoors. In the United States, homes were often crowded with many family members trying to integrate themselves into U.S. society. Because U.S. homes were constructed to be much more air-tight than the houses they were used to, their cooking habits caused a number of fatal fires and, in some cases, fire-related deaths even though a fire did not break out. These small charcoal cooking fires tended to use up the available oxygen in the home or apartment, and they also gave off deadly carbon monoxide gas. Many Asian immigrants were killed in such fires—not because they were doing anything wrong, but because the norms of behavior they had learned in their former countries did not translate into the setting of their new home.

Of course, these types of fires are not limited solely to Asian immigrants. This is only an illustration of how belief, and behaviors, can affect fire risk. As the Asian immigrant population integrated themselves into U.S. society, these types of cooking fires dropped off and are far less a factor in fire risk today. Yet kitchen fires of all types remain one of the principal causes of fire in the United States—with the young (unsupervised) and elderly making up the target risk audiences.

People also have beliefs and demonstrate behaviors toward fire that are not ethnically related. For example, religious beliefs about the use of candles can span national cultural identities and can be a factor in fire risk based on a variety of approaches to religion or spirituality.

Many people use candles extensively in religious ceremonies, with no adverse affects. But we often see fires caused by candles that have been left unattended, which have ignited combustibles nearby, destroying homes and businesses and killing or injuring people. Many people also use candles a great deal in their homes not as part of an organized religious rite, but rather as a more generalized spiritual belief in the power of fire and light. Even if it is not a specific religious belief, this is still a collective value and behavior that can increase risks from fires.

Again, it is not accurate to say that all candle fires are related to religious beliefs or values. In fact, where alcohol or drugs are contributing factors, unattended candles may have nothing to do with religion or spirituality. The point is that sometimes a set of values or beliefs, and accompanying behaviors, may increase fire risk. Moreover, those behaviors may be motivated by value systems that are not ethnically driven.

There is another distinction that should be drawn here. The elderly make up a good percentage of fall injuries, but this is not because of their values or beliefs. It is more a matter of age, and the gradual failing of their bodies and balance systems. The elderly are more susceptible to damage from a fall because of their aging bodies, and recovery times are longer and more expensive. In fact, falls can

Some cultures value candles more than others—and their use can create more fire risk, especially for those unfamiliar with how they might react to common combustibles. (*Courtesy of Vancouver Fire Department, WA*)

be deadly. Therefore, the link between a certain population and a particular risk is not always driven by cultural values.

## CULTURE AND DECREASED FIRE RISK

It is also possible for cultural values to contribute to a decreased risk of fires. International studies—notably by TriData Corporation, a division of System Planning Corporation of Virginia—have drawn some links between various cultures and reduced risk of fire. The Japanese culture is often given as a primary example.

In Japan, a history of fires, combined with relatively flammable housing construction and the tight spacing between houses, has created a cultural value that focuses on fire safety. According to Phil Schaenman of TriData, the Japanese put much more emphasis on fire prevention than do fire departments in the West. But the cultural view of fire is also radically different than the United States. In our nation, we tend to view unintentional fires as "accidents"—meaning they are beyond our control—and the victims of fires are treated with compassion and support.

In Japan, people who have fires are admonished by their neighbors for being careless. In one case that Mr. Schaenman related, a family who had an unintentional fire was actually driven to move from their neighborhood by angry neighbors who had been threatened by the fire. It did not matter that it was unintentional. In Japan, a fire is viewed as a more controllable event than in this country, and it is a matter of public shame to have one. That value is driven by the fact that a fire started by one person can be a threat to many others who live very close by.

Anecdotally, we are beginning to see such a value change in the United States. In multifamily properties where people live close together, a fire caused (whether intentionally or not) in one unit can affect many others in the same complex. A fire that occurs in a first-floor apartment can drive families who live on the second and third floors out of their homes through no fault of their own. Families without renter's insurance for their belongings can lose everything they own in a fire caused by someone else.

Aside from renter's insurance, people who live in multifamily environments are beginning to realize that a careless act can be controlled. Fire prevention peers around the United States are beginning to note some social pressure for those who have unintentional fires to take responsibility for their actions, and even to compensate those who have lost belongings through no fault of their own.

One such case in Vancouver, Washington, arose from juveniles playing with fire. The fire spread, damaging some nearby apartments. The owner of the damaged apartment building, having no relation to the youths who had set the fire, pressed officials to punish the parents of the youths and to provide information (despite juvenile confidentiality issues) so that the parents could be held financially responsible for the actions of their children.

One case is hardly a scientific study, but peers around the nation are beginning to see that values in the United States are becoming more similar to those in Japanese and other cultures, where personal responsibility for the results of fires is more likely to be assigned to those who cause them through some careless act.

According to Mr. Schaenman, similar cultural values about personal responsibility exist in many European nations as well. These nations have even acted on the notion of personal responsibility in some cases by enacting laws that prohibit

The Japanese have placed a greater emphasis on fire prevention than U.S. fire departments, devoting much higher percentages of overall fire protection budgets to fire prevention efforts. (*Courtesy of Tri-Data, a Division of System Planning Corporation*)

full insurance coverage of personal property. Thus, everyone has some financial—as well as personal—interest in preventing unintentional fires from occurring in the first place.

## HISTORY'S IMPACT ON CULTURE

In cultures where those who start fires even "accidentally" are held responsible for the results, fire loss rates tend to be much lower than in the United States. There are no scientific studies to substantiate this supposition, but it is widely believed that the history of fires in European and Japanese cultures contributes greatly to the cultural emphasis they place on fire prevention. According to some experts, including Phil Schaenman of TriData, the early fires in the United Kingdom (see the discussion of the Great Fire of London in Chapter 8) and the fires caused by wars (such as the bombing of London in World War II) have helped create a cultural value for fire safety. The construction materials and methods used in the United Kingdom are sometimes greatly different from those used in the United States, but their view of fire protection is notably different as well. Those differences are driven by a long history with fire, its devastation, and an understanding that individuals can control it.

Those looking for a deeper understanding of the role that history plays in developing a more fire safe culture would be well served to examine the international studies conducted by TriData and the Centers for Disease Control and Prevention.

# The United States and a Multicultural Society

The culture in the United States is changing, due in large part to the continuing immigration that brings a constant flow of new cultures into the so-called melting pot of U.S. society.

One (unscientific) source (http://en.wikipedia.org/wiki/Multiculturalism#United_States) notes that the society or culture of the United States has been predominantly Western, with its own unique characteristics and developments such as dialect, music, arts, and cuisine. Today the United States of America is a more diverse and multicultural country as a result of mass immigration from many countries throughout its history.[1]

The article notes that the early influence of British culture helped shape the culture of the United States, including the spread of the English language and legal system. However, other European immigrants brought with them values, behaviors, and traditions that also shaped American culture.

More recent immigration from Latin America, Asia, India, and other countries has begun to create numerous subcultures, leading some to describe the United States as a melting pot of human society. Others, however, view it as more of a mixed salad, with distinct flavors that can enhance one another collectively but retain their own individual tastes. This may be a poor metaphor, but it does describe the trend toward a balance of one collective culture for the nation while various subgroups attempt to maintain cultures and values of their own.

According to data kept at the Center for Public Education, the U.S. population is growing older, with "baby boomers" (those born in the years after World War II) comprising a significant portion of the population. That aging population is also becoming much more diverse than it has been in the past, with significant increases in Hispanic and Asian/Pacific Islander populations. Though the numbers do not show up in national databases, an increase in Eastern European populations, although relatively small, also has an impact on cultural values in the United States.

The changes in these demographics will not affect all states to the same degree. The nation's population density has changed, with more than half living in just 10 states (California, Texas, New York, Florida, Illinois, Pennsylvania, Ohio, Michigan, New Jersey, and Georgia).

The U.S. Census Bureau provides a wealth of information on demographic trends for each part of the nation. Keeping track of those changes, and the cultural changes accompanying them, will be critical for fire prevention managers in the coming decades. Most of the public attention devoted to these changes centers around basic Social Security and Medicare funds. But there will be direct impacts on the values and societal views of fire safety as a result of the changing culture in the United States—such as the fact that many will not be able to read English, and some will not be able to read in any language at all.

Each demographic population or cultural base has unique attributes, some of which will affect fire prevention efforts aimed at them. It seems reasonable to assume that larger populations of these cultures will mean they have a larger share of the nations' fire problem as well. Consequently, reaching them with our prevention messages will become even more important than it is today.

What follows are some brief anecdotal examples of the cultural values of these populations, along with some recommended resources for learning more about them.

## HISPANIC POPULATIONS

According to a report prepared for Ohio State University,[1] the term *Hispanic* was created by the federal government in the early 1970s to describe a large and diverse population with a connection to the Spanish language or culture from Spanish-speaking countries. The term *Latino* is gaining popularity, as it reflects the origin of the populations, from Latin America.[2]

In that report, Ann W. Clutter and Ruben D. Nieto describe aspects of Latino culture that can influence the way fire prevention practitioners reach them. Among them is the fact that family is traditionally the focal point of social activity. Family ties are strong, with large families often living together in relatively small spaces. The preservation of the Spanish language is a strong social norm.

According to the report, Hispanics usually assign great importance to looks and appearance as a matter of honor and pride. Hispanics tend to be more relaxed about time and punctuality than other populations in the United States; it is socially acceptable to show up late.

Though Hispanic populations are oriented more toward public education than some others, the report outlines the importance of actively interacting with this group in order to reach them effectively. The implications are evident for public fire prevention and life safety efforts.

According to Nutter and Nieto, gaining the trust of these audiences is a critical factor in reaching them. Involving Hispanic community leaders in planning and delivery of programs is a way to gain their trust and support.

The authors point out that churches, local libraries, and recreational centers (with child-care arrangements, if needed) may be appropriate places to hold educational programs with Hispanic audiences, and because information is passed mostly by word of mouth, grocery stores and churches may be a good place to meet, visit, and exchange information.

## ASIAN/PACIFIC ISLANDER POPULATIONS

Asian populations make up about 5% of the U.S. population (according to Asian Nation) but are one of the fastest growing segments of our society. Asian Nation maintains a Web site that provides a good deal of insight into the Asian culture. That organization defines "Asian" as a group living in the United States who self-identify as Asian or Pacific Islander ancestry. That may be the case whether they are immigrants or born in the United States.

Asian Nation acknowledges that there is a debate over how to refer to certain Asian groups—Filipinos or Pilipinos, Koreans or Coreans. The Smithsonian maintains a Web site that has a wealth of cultural information, including a section devoted to Asian Pacific Americans.

There is a long history of Asian populations in the United States—and not all of it has been positive for those populations. For Chinese Americans in particular, information located on the Smithsonian's Web site pays homage to the sweat shops of the past, and the near slave labor camps that influence their experience and their views of U.S. culture.

According to an article posted on the Web site for the Washington State Commission on Asian Pacific American Affairs, the Asian Pacific Islander population in the United States is diverse in its own right and complex. Gary Huang, of Teachers College, Columbia University, makes several points about this particular culture.[3]

He stipulates that there are really three distinct groups among the Asian/Pacific Islanders. The Pacific Islanders are made up of the island populations of Hawaiians, Samoans, and Guamanians. There is also another group from Southeast Asia, such as the Vietnamese, Thais, Cambodians, and other Indochinese populations. Finally, there are the east Asians, made up mostly of Chinese, Japanese and Koreans.

He stresses I, "It is important not to generalize an understanding of one group to another. . . . The Vietnamese, . . . have a sophisticated literate culture and strong abilities to adapt to the market society; the Hmong have no written language, nor skills that are easily applicable to American labor needs."

The implications for fire safety educators should be evident. A more thorough understanding of the various subcultures that make up the Asian and Pacific Islander population in the United States is necessary for reaching these groups in any prevention setting. For example, dealing with restaurant owners of API heritage may be daunting, and in some cases may require interpreters. Similarly, conducting fire investigations for these and other populations might produce an entirely different approach—with an understanding of their culture—than was previously thought important by fire prevention managers. In the same article, Mr. Huang also noted in the same article that Asians may unconsciously favor ambiguity in their communication to avoid giving offense. Repeated head nodding and lack of eye contact (as an example) could give fire investigators the wrong impression about truthfulness. And if English is a second language, there may be even more difficulties lost in translation. All of these issues and more present challenges to fire prevention practitioners.

## EASTERN EUROPEAN POPULATIONS

Like Asian/Pacific Islander populations, Eastern European populations are diverse. They are not all Russian, as many assume. Among the Eastern European cultures that have immigrated, the variety may include Latvian, Bulgarian, Croatian, Lithuanian, Polish, Czech, or Hungarian populations. The differences between these groups may be important to understand, even though some similarities exist.

Because of Soviet-era politics or long-standing cultural imprinting, Eastern European populations may have a much different view of authority than we have in the United States. For example, they may have a suspicious approach to government intervention, and therefore a tendency to try and figure out what motivates certain questions. This can have a direct impact on a fire investigation. When a government official is asking questions about a fire's origin, and a person from the former Soviet bloc is answering elusively, the usual conclusion about suspicious motivation on the part of that person may be very much mistaken. People from those cultures may instead be thinking of what might be behind the question of the fire official, and in order to protect themselves, answer elusively to avoid trouble from the government—even when an unintentional fire would cause no repercussions for that individual.

It is an oversimplification to sum up Eastern European culture in this fashion, and those interested in a more in-depth understanding may find some valuable resources at the University of North Carolina Center for Slavic, Eurasian and East European Studies.[4] However, the point once again is that people of another culture may respond to the needs of the fire prevention community in a very different fashion than that of which we may be familiar.

## AFRICAN AMERICAN POPULATIONS

To understand African American culture, it is necessary to understand African American history. Many African Americans can trace their family heritage back only a few generations to slavery in the United States, and personal accounts of active segregation programs are remembered firsthand by people who are alive and relatively young today. That dynamic has shaped a culture that is unique in the general U.S. population—so that even those who have lived in the United States all their lives may have values, beliefs, and traditions that are different from of the majority of the population.

African Americans for years represented a segment of the U.S. population that had a higher than average percentage of fires, fire injuries, and deaths than the nation as a whole. They still represent a high-risk group for fire risk assessments, but recent studies by the National Fire Protection Association have drawn a stronger correlation between fire loss rates and income levels than factors such as ethnicity or national origin. In short, any population with more members in or near poverty would be at greater risk of fire.

There are likely more reliable sources of information on African American culture, but a good summation exists on Wikipedia.[5]

The article stipulates that there are cultural values of African American behaviors and attitudes that reflect the fact that they are a distinct culture within the "melting pot" of American society. For example, the article outlines that because slavery restricted their ability to learn to read or write, the communication patterns among African Americans took on a specific aspect. Because slaveholders prohibited education as a way to control slave populations, oral traditions became the primary means of preserving African American history, morals, and other cultural values. Storytelling became a way to pass on traditions from one generation to the next.

The legacy of that oral tradition displays itself in many aspects of modern African American culture. According to the article and some personal accounts, African American preachers may perform a story, rather than just speaking it. The emotion of the subject is brought about by speaking tone, volume, and movement. Often song, dance, verse, and structured pauses are used to emphasize emotional elements of a sermon.

And unlike more dominant (at least during current times) traditional American or Western cultures, it is an acceptable and common practice in the African American culture to interrupt and affirm a speaker.

The oral tradition is just one aspect of African American culture that establishes them as a distinct population and culture within the United States. The article proposes that African American culture developed separately from the predominant American culture because of many years of racial discrimination and segregation—even after slavery was abolished.

A wealth of information about African American history and culture is available on the Internet. It is important for fire prevention managers to realize that when dealing with African Americans, they may in fact be dealing with a distinct culture and its influences on a specific U.S. population.

Dealing with different cultures or subcultures is a complex issue, and getting expert help and establishing relationships within various minority communities is highly recommended. The work of Fire 20/20 is notable in this regard. Readers can gain more information on the topic via the Fire 20/20 Web site. In any event, fire prevention managers need to understand major U.S. cultures to target their fire prevention efforts appropriately.

# The Fire Service Culture and Its Impact on Fire Prevention Efforts

Like specific ethnic or national cultures, organizations tend to develop local cultures of their own. The basics for defining culture in this context are still the same. Organizational culture is defined as the norms of behavior and values that an organization finds important, and for which those exhibiting these cultural traits (norms and behaviors) find acceptance by the group.

In an article for an online management library Carter McNamara refers to organizational culture as the personality of a company.[6] A simplistic example may be identified by determining the value the organization and its members place on time. Do they value being to work on time, or is it a more relaxed atmosphere? If the value is to show up on time—typical for the fire service—then those who show up late for whatever reason will tend to be ostracized by the group. A particular value may in fact be important to an organization for a very good reason. For example, when firefighters depend on those who replace them on duty to be reliable and on time so that they can meet family or other obligations, an organizational value that places importance on being at work on time is driven by necessity. But these values may not take the form of written rules; they may be something that is learned and regulated informally.

So how do we define the organizational culture of the fire service? Each fire department may have unique aspects of "personality." But anecdotal examples from a variety of fire departments suggest that some aspects of the culture in the fire service are common throughout the United States—and even in other nations. One example is the value of teamwork, because firefighters must function as a team. When team members rely on each other for critical job functions, and to protect each other from harm, then teamwork becomes a very important value of an organizational culture. Teamwork defines a norm of behavior, and those who do not exhibit that trait are at risk of being shunned by the organization. How long would we expect a firefighter to last who tended to operate independently from the group? While others were combining resources to successfully attack a fire, a lone member takes off on his or her own. How long would the group tolerate such behavior?

This is another simple example, of course, but it expresses itself in far more subtle ways in the organizational culture. Even if someone does their job well—and functions well as part of a firefighting team—they could run afoul of organizational cultural values. Many firefighters function so tightly as a team that they

The organizational culture of the fire service in the United States places value on teamwork, strength, courage, and heroism—most often driven by the needs of the job. (*Courtesy of Spokane Fire Department, WA*)

bond together in team functions outside of the workplace. If that is part of the culture of the organization, how long will a member who does not participate in these functions outside the workplace last?

This informal aspect of organizational culture can be a particular problem for those who come from different cultures and who do not exhibit the same team attitude beyond the workplace.

The point here is that organizational cultural values can span different fire departments—and although they may originate by necessity, they can also become a problem for a multicultural or dual-gender fire service. Those who do not adapt to all aspects of teamwork—formal and informal—may have difficulty adapting to the organizational culture and being "accepted" as members of the organization.

Other aspects of fire service culture are also driven by the reality of the job. Because the work is physically demanding—and dangerous—strength and courage are traits valued by the group. Again anecdotally, the value for heroism is reflected by the importance the fire service places on it in fire departments all across the nation. This observation is illustrated by funeral processions for those killed in the line of duty (firefighting) all across the nation. The fire service comes together to honor those who died while risking their lives to save others.

The National Fallen Firefighters Foundation was formed as a recognition of this aspect of fire service culture. Although those involved in the NFFF recognize the value of honoring those who have been lost in the line of duty, they also recognize that the same organizational culture that values heroism can also value risk taking. And those who value risk taking may be difficult targets for safety messages.

## FIRE PREVENTION AND FIRE SERVICE CULTURE

How, then, is fire prevention valued within the fire service? Is it as important as teamwork, risk taking, physical strength, or heroism? Generally speaking, it is not. The fire service consists in large part of firefighters, emergency responders, and fire medics (firefighter or paramedics) who make up the bulk of a fire department's personnel resources. Prevention usually accounts for less than 7 or 8 percent of a fire department's personnel. The tendency to place value on what is most prevalent is common.

Throughout the history of the fire service in the United States, fire prevention has been espoused by many fire service leaders and fire chiefs as the real reason for existing. Some have been known to say that a fire is nothing more than an indicator that prevention efforts were not successful. But the bulk of public resources still go toward emergency response (often medical-related as a majority) and firefighting. So valuing fire prevention with something other than dollars has become the norm in the U.S. Fire Service.

Some fire departments go against this basic concept. They value prevention and make it part of the job of everyone in the organization—not just those assigned to that particular job function. Dallas, Texas, for example, does a great deal of prevention work (home visits, smoke alarm installations) with on-duty fire crews. Others, such as Vancouver, Washington, are taking proactive steps to make station-based prevention activities a normal staple of the job.

It is difficult to take an organizational culture that values risk taking, heroism, and strength and to instill it with the values of prevention. Doing so requires a great deal of organized effort. According to an article on About.com: Human Resources, by Susan Heathfield,[7] changing organizational culture must start at the top. The executive leadership in the organization must embrace the values they desire to have adopted by the organization. They must model those values and communicate them clearly and often—and the values must be rewarded.

In the fire service context, when prevention is supported by the organization's leadership—and when promotion decisions are based not just on firefighting skill but on a commitment to prevention—then the organizational values will begin to change, and prevention will begin to be valued more than it has been in the U.S. Fire Service.

## Cultural Solutions for Fire Prevention

The fact that the United States is becoming an increasingly multicultural society has already been mentioned as a significant challenge for prevention efforts. However, one organization has found value in reaching out to different cultures on behalf of the fire service, and enlisting their support in fire prevention efforts.

Larry Sagen, the executive director for Fire 20/20, has been involved for many years in getting fire departments to reach out to more diverse audiences. But what he and his organization have discovered is that in reaching out for recruitment efforts, they have established relationships with members of minority communities who have a desire to help their own community. Faced with challenges of fire prevention, many of these people have organized themselves to help local fire prevention efforts. They have provided access to minority populations that a fire department would not ordinarily have—and this access exists because of efforts by local fire department to reach out to them.

The goal of Fire 20/20 is to increase diversity in the fire service workforce, and to help the fire service understand the value of that diversity. One concrete value that has arisen out of those efforts is a commitment on the part of different cultural groups to support fire prevention efforts.

**diversity and fire prevention**
■ taking different languages and culture into account, and reaching out to others to create diverse partnerships that aid fire prevention efforts

The work of Fire 20/20 has produced a natural understanding of **diversity and fire prevention**. The issues are closely related on several levels.

It is important to keep in mind that educational or outreach efforts must be designed to reach a diverse audience. It is also important to recognize that when communities look at anyone trying to reach them, they will open themselves up more readily to people who look and act like them.

## Generational Culture

**generational culture**
■ the differences in values, beliefs, and norms of behavior that exist between differing age groups

**Generational culture** can be a significant factor affecting prevention efforts. Separate from more typical racial, ethnic or social cultural descriptors, those involved in fire prevention should understand that differences may be profound among people of different ages.

The Web site Generation Culture is tracking differences between what they describe as Matures, Baby Boomers, Generation X'ers, and Nexters. A variety of material exists on the Internet about the traits each generation exhibits in general terms. For example, Generation Culture discusses the differing values between Matures and Baby Boomers on retirement. Generally, they are finding that Baby Boomers plan to work far longer—perhaps never truly retiring at all. Articles on the Web site also examine the ramifications of the use of technology for the younger generation—and their social networking behaviors. It is not uncommon to see younger people using their cell phones and the Internet to network with a *very* broad spectrum of friends and acquaintances by texting each other and posting messages and photos on sites such as Facebook.

These are just examples of how the differences between age groups may affect our prevention efforts, and how we try to appeal to audiences of different ages.

## Summary

There are traditional cultural beliefs in the United States that influence fire prevention. Because our history with fire has relied on insurance to cover losses, the culture in the United States is less oriented toward prevention than other nations, particularly in Europe and Japan. Because of our increasingly multicultural society, fire incident rates are influenced by a more diverse set of cultural values, which sometimes can increase fire risks and sometimes decrease them.

That same evolution toward a more multicultural society in the United States presents significant challenges to fire prevention efforts, including audiences that do not respond to traditional educational outlets. And because the cultural values may be significantly different among all these groups, even reaching these audiences in their own language may not work because we have not established the proper relationships with them.

Establishing those relationships and reaching out to the various cultures in our own areas can have positive impacts on our prevention efforts. We sometimes find willing partners who are anxious to act on behalf of their own safety, and that of their community, by carrying fire prevention messages to places we would not normally reach.

The fire service has an organizational culture of its own, which has common elements throughout the United States. Fire department personnel usually value teamwork, strength, courage, and heroism because of the dangers of the job. More fire departments are finding value in promoting prevention measures—as a way to reduce risk for firefighters themselves, but also as a way to provide a full range of protection measures for the communities they serve.

## Case Study

As the newly appointed fire marshal for your jurisdiction, you have been tasked with elevating the fire prevention efforts for your community. In visiting the local fire stations, you have noted that the previous fire marshal was not well respected—and that the prevention bureau was viewed as a group of oddballs who really did not fit into the organization. One complaint commonly heard was about significant delays in response by fire investigators, which in turn kept emergency responders on scene for long periods of time waiting for overhaul so that investigators could view the fire scene before it was torn apart. Lately, firefighters have been starting overhaul efforts on their own—and fire operations managers have been reluctant to stop them because of a heavy workload.

Training for prevention personnel has been cut severely in recent years, and the morale in the fire marshal's office is demonstrably low. The leadership of the organization has not viewed prevention in a positive light, in part because of disagreements over safety conditions between the previous leaders in that office and the union. The budget for the fire marshal's office is actually larger than you were led to believe, and although it is not everything you might want, there is some room to maneuver.

This community has an increasing Hispanic population, with some Russian immigrants forming a small but significant portion of the community (and the voting public).

# Case Study Review Questions

1. To what extent does the firefighters' opinion of the prevention bureau determine its image among department leaders?
2. Is there a link between what the firefighters believe, and real-world demands on them?
3. What efforts should be made—and how would you go about making them—to increase outreach efforts in your community?
4. Explain how increasing the training budget might affect the image of the fire marshal's office in the community—and with fire department leaders.
5. How may U.S. cultural beliefs affect fire prevention efforts?
6. How are cultural beliefs about fire safety in different nations and cultures different from those in the United States?
7. How will the fact that the United States is becoming more multicultural affect fire prevention efforts?
8. How can multicultural advances in the United States be of benefit to prevention efforts?
9. Describe aspects of Russian-speaking cultures and immigrants that may be an impediment to fire investigation activities.
10. Describe elements of organizational culture.

PEARSON
## myfirekit ™

For additional review and practice tests, visit **www.bradybooks.com** and click on MyBradyKit to access book-specific resources for this text! Your instructor may also assign Additional Project work related to topics in this chapter.

Register your access code from the front of our book by going to **www.bradybooks.com** and selecting the mykit links. If the code has already been scratched off, go to **www.bradybooks.com** and follow the MyBradyKit link from there.

# Endnotes

1. N.A. (n.d.). Multiculturalism. Retrieved from http://en.wikipedia.org/wiki/Multiculturalism.
2. Clutter, Ann W., and Ruben D. Niety. (n.d.). "Understanding the Hispanic Culture." Ohio State University Fact Sheet. Available at http://ohioline.osu.edu/hyg-fact/5000/5237.html.
3. Huang, Gary. (n.d.). Retrieved from www.capaa.wa.gov/about/beyondCulture .pdf.
4. Center for Slavic, Eurasian and East European Studies, UNC-Chapel Hill. (February 1997). Retrieved from http://www.unc.edu/depts/slavic/publications/brochure2.html.
5. N.A. (n.d.) "African American Culture." Retrieved from http://en.wikipedia.org/wiki/African_American_culture.
6. McNamara, C. (2000). *Organizational Culture*. Authenticity Consulting, LLC. [Adapted from *The Field Guide to Leadership and Supervision*.] Retrieved from http://www.managementhelp.org/org_thry/culture/culture.htm.
7. Heathfield, S. (n.d.). *How to change your culture: organizational culture change*. Retrieved from http://humanresources.about.com/od/organizationalculture/a/culture_change.htm.

# 10

# Economic Influences on Fire Prevention

## OBJECTIVES

After reading this chapter you should be able to:

- Identify the relationship between economic factors and fire prevention.
- Identify the difference between direct and indirect fire losses.
- Describe the cost of fire in the United States.
- Describe the disparity between high costs of fire and lower perceptions of fire risk.
- Describe policies and programs that afford economic trade-offs for fire prevention.

# Economic Influence

What do we mean when we talk about economic influences and fire prevention? It is a bit more complex than just saying that fire costs something in terms of economic loss. But the relationship between money and fire loss is perhaps the most compelling factor in addressing it through prevention. As Benjamin Franklin said, "An ounce of prevention is worth a pound of cure." Prevention of fire loss is a more cost-effective way of dealing with the fire problem—though it is difficult to quantify results. More is said about evaluation of fire prevention programs in Chapter 12. But intuitively, if nothing else, most people realize that preventing a fire from occurring is much cheaper than putting one out and suffering the losses it has produced.

The relationship between economic factors and fire prevention is part of the overall economic picture related to fire protection costs as a whole. The following excerpt is from a report produced for Forest Encyclopedia Network that provides some excellent insight into the relationship between the economy and fire protection[1]:

> If Robinson Crusoe lived alone on an island that was prone to wildfires, how would he deal with wildfire risk? Because of the wildfire risks, Crusoe has an incentive to spend effort and time to reduce the risk of losing his shelter to fire. Time and effort spent clearing brush and trees around his shelter is time that he would use to gather water or hunt and fish for food. But Crusoe does so because he expects this pre-fire effort to sufficiently reduce the effort needed to fight a wildfire and the risk of having to rebuild when a fire does come along. How much effort should he exert clearing vegetation? He would likely start by clearing the vegetation that poses the most risk—perhaps bushes right beside the shelter—and might work incrementally away from his shelter. With every concentric ring outward, the additional benefit to him in terms of expected suppression and rebuilding effort is likely to diminish, and Crusoe would likely stop clearing at the point that the benefit from one additional concentric meter of clearing is outweighed by the cost of lost hunting and fishing time.

This example of the "Robinson Crusoe Economy" is not limited to wildfire protection. It is a great illustration of how people will make decisions based on the level of value they perceive prevention efforts will afford them.

## HIERARCHY OF NEEDS

Abraham Maslow postulated that all humans are driven by unmet needs—and that the higher-level needs act only as motivators once the lower-level needs have been met. For example, the lowest level of need is for basic physiological requirements, such as food, shelter, and water. After these basic needs are met, people may be motivated to achieve higher levels such as safety and security, leading all the way to a higher state of being (self-actualization) where they are motivated by unselfish desire to help others. Readers can learn more about Maslow's **hierarchy of needs** at the Abraham Maslow website.

**hierarchy of needs**
■ the theory produced by Abraham Maslow in 1943 that describes a tiered pyramid of needs that all humans have

The important point about Maslow's theories is that there are basic needs even more important than safety and security, and until they are met people will not pay adequate attention to their own (or family members') safety. For those who have not met their most basic needs, putting food on the table will be more important than, for example, purchasing a smoke alarm. Consequently, money—which provides for basic needs—is a critical issue individually and collectively.

Another aspect of the economy also affects fire protection and fire prevention specifically. Insurance industry studies show the relationship between a bad economy and increased fires. It is common knowledge among fire professionals that insurance fraud may increase when people are not able to make ends meet, and some in the insurance industry have pointed out that in today's economy insurance fraud has reached very high levels.[2]

Anecdotal evidence from other fire chiefs and fire marshals from around the nation has supported the opinion that a slow economy may stimulate arson fires for insurance fraud. It is not a scientific study—but clearly the economy affects fire rates and fire prevention in many ways.

The issues are complex, but for the purposes of this chapter, when we refer to *economic influences* on fire prevention, we are talking about the relationship between money and fire, or more specifically fire prevention. That would include examining direct and indirect losses from fire, the relationship that fire losses have with overall economic health, the relationship between costs of fire and perceptions of risk, and the economic value of fire prevention.

## Direct Versus Indirect Costs of Fire

**direct fire losses**
■ the loss of fire in dollars

**Direct fire losses** are fairly easy to describe, but more difficult to assess. When we consider the cost of a fire in dollars, it is fairly simple to state the loss in those terms. The figure is typically arrived at by estimating the cost of replacing whatever the fire destroyed. Examples include estimating the value of a building and its contents or the timber value of wildland lost to fire. However, things are rarely that simple or straightforward, and describing the cost of fire and fire losses is no different.

**indirect fire losses**
■ those that can be extrapolated from the direct costs associated with replacement

**Indirect fire losses** are those that can be extrapolated from the direct costs associated with replacement, but are more difficult to describe. The cost of a human life, for example, cannot truly be quantified, though some have tried to describe it in terms of earning power lost over a lifetime. The loss of a loved one, and the impact of such a loss on a family or community, cannot be described in such simple terms.

Similarly, the indirect costs associated with repairing the physical damage caused by fire may run far higher than the portion of a building or property actually destroyed. Consider these following examples of indirect loss:

- The loss of employee payroll while the firm is closed
- The loss of business to competitors while a firm is closed
- The loss of business to suppliers who provided goods and services to a company closed down by fire
- The loss to a community in terms of both property and sales tax from a business that is closed
- The cost of temporary housing for those whose home was destroyed by fire
- The loss to related businesses that rely on the employees from a closed business for their economic livelihood—such as the food vendor that supplies a manufacturing company for their employee breaks and lunches
- The timber jobs that were lost because a forest was destroyed by fire
- The expense associated with increased feed costs for cattle that used to graze on inexpensive public land that was destroyed in a large natural cover fire

These are examples of indirect cost that can be calculated in some fashion, but not exactly, yet are still part of the economic losses that a fire can produce. But when we consider some **intangible fire losses**, our ability to describe them becomes less about money and more about emotion.

Intangible losses are those we suffer when we cannot describe the loss in financial terms. The loss of a loved one has already been mentioned. Even if a dollar amount cannot be set, everyone can relate to the emotional pain a fire may produce. And what about the loss of habitat for other species? If the species in question is endangered, how can we set a dollar amount to describe the possible loss of an entire species because their habitat was destroyed by fire? That can be a two-sided coin, of course—fire is part of the rhythm of nature, and some species thrive in the aftermath of a fire in a natural habitat.

Still, how could we begin to describe the loss of a natural treasure, such as a large recreational area that is destroyed by fire? What is the cost to the people who frequented that area but are no longer able to enjoy it? Or even more poignant—how would we describe the loss of a historic building that means much more to a community than the cost of its replacement—such as a church where family members worshipped, were married, had funeral services, and bonded over generations?

These combined factors begin to describe the loss we can suffer when an undesirable fire occurs. And they are typically what we use to describe how much fire costs our nation. But there are other costs associated with fire we must consider when we are describing the overall burden of fire on a society.

**intangible fire losses**
▪ those we suffer when we cannot describe the loss in financial terms

# The Cost of Fire in the United States

Both the U.S. Fire Administration and the National Fire Protection Association (NFPA) produce reports estimating the cost of fire in the United States. In addition to economic losses, the NFPA model has an aspect that is not usually considered: the cost of fire protection.

According to statistics in 2004 (the most recent available for such an estimate), the NFPA estimated the overall cost of fire in the United States at somewhere between $231 billion and $278 billion—or about 2.5% of the gross domestic product of the United States at that time. Those figures were arrived at by estimating the direct and indirect losses associated with fire—but also included the cost of both fixed and mobile fire protection in the nation. The NFPA extrapolated cost estimates for fire insurance and fixed protection (e.g., sprinklers, alarms, building compartmentalization), as well as the cost of paid fire departments. The range of costs was largely accounted for in estimating the value of volunteer fire departments in the United States.

From this point of view, the true cost of fire in the nation shows compellingly how much of our national economy is tied up in fire protection—which implies that we should be looking at more cost-effective ways of managing the fire problem.

As an example of how to look at the cost of fire, this approach is very different from what has occurred in the past. And when we are looking for examples of how to quantify the losses—and to describe them in a meaningful manner—we can look to both the U.S. Fire Administration and the National Fire Protection Association for ways to do so.

# Fires in the United States During 2008

**1,451,500** fires were reported in the U.S. during 2008.

- down **7%** from 2007
- **3,320** civilian fire deaths
- **16,705** civilian fire injuries
- **$15.5 billion** in property damage
- **103** firefighter deaths

 Firefighter deaths are not restricted to fires.

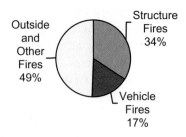

**Fires in the United States During 2008**

Outside and Other Fires 49%

Structure Fires 34%

Vehicle Fires 17%

**515,000** structure fires occurred in the U.S. during 2008.

- decrease **3%** from 2007
- **2,900** civilian fire deaths
- **14,960** civilian fire injuries
- **$12.4 billion** in property damage

**236,000** vehicle fires occurred in the U.S. during 2008.

- down **9%** from 2007
- **365** civilian fire deaths
- **1,065** civilian fire injuries
- **$1.5 billion** in property damage

**700,500** outside and other fires occurred in the U.S. during 2008.

- down **9%** from 2007
- **55** civilian fire deaths
- **680** civilian fire injuries
- **$0.2 billion** in property damage

*The California Wildfires 2008, with an estimated property damage of $1.4 billion is included in the total property damage but it is excluded from the property damage estimates for structure fires, vehicle fires and outside and other fires.

Source: *Fire Loss in the United States During 2008* by Michael J. Karter, Jr., NFPA, Quincy, MA
*Firefighter Fatalities in the United States – 2008*, by Rita F. Fahy, Paul R. LeBlanc and Joseph L. Molis, NFPA, Quincy, MA

The true cost of fire in the United States includes not only the direct and indirect losses, but also all the costs associated with protecting against fire. That includes fixed protection systems built into buildings, as well as the estimated costs of paid and volunteer fire departments. (*Reprinted with permission from National Fire Protection Association, © 2008 NFPA*)

# 2008 United States Fire Loss Clock

Every **22 seconds** a fire department responded to a fire.

One vehicle fire was reported every **134 seconds**.

One outside fire was reported every **45 seconds**.

One structure fire was reported every **61 seconds**.

One civilian fire injury was reported every **31 minutes**.

One home structure fire was reported every **82 seconds**.

One civilian fire death occurred every **2 hours and 38 minutes**.

Source: *Fire Loss in the United States During 2008*, by Michael J. Karter, Jr., NFPA, Quincy, MA, August 2009

The way that NFPA describes fire losses provides a good example of how to prepare reports on local statistics. (*Reprinted with permission from National Fire Protection Association, © 2008 NFPA*)

Gathering the data on loss is just the first step in making our case for describing its impact. How we present the information is just as important. For example, consider the latest statistics from the NFPA.[3]

In 2007, there were 1,557,500 fires reported in the United States (down 5% from 2006). These fires caused 3,430 civilian deaths, 17,675 civilian injuries, and $14.6 billion in property damage.

- 530,500 were structure fires (up 1% from 2006), causing 3,000 civilian deaths, 15,350 civilian injuries, and $10.6 billion in property damage.
- 258,000 were vehicle fires (down 7% from 2006), causing 385 civilian fire deaths, 1,675 civilian fire injuries, and $1.4 billion in property damage.
- 769,000 were outside and other fires (down 9% from 2006), causing 45 civilian fire deaths, 650 civilian fire injuries, and $0.8 billion in property damage.

Also consider the 2007 U.S. fire loss clock[4]:

- A fire department responded to a fire every 20 seconds.
- One structure fire was reported every 59 seconds.
- One home structure fire was reported every 79 seconds.
- One civilian fire injury was reported every 30 minutes.
- One civilian fire death occurred every 2 hours and 33 minutes.
- One outside fire was reported every 41 seconds.
- One vehicle fire was reported every 122 seconds.

This way of describing the nature of the fire problem goes beyond the direct and indirect costs, and begins to make clear some of the intangible ways fire loss affects our nation. When combined with some simple charts or graphs, we not only get a more full picture of how the fire problem affects our nation, we get examples of how to construct reports for our own jurisdictions.

But sometimes, the most compelling example of fire loss is a personal story told by a member of the community who has suffered from it. In such circumstances the economic impact might be slight, but a high-visibility personal loss, and the political effect produced by that loss, might be enough to sway financial decision makers to support prevention efforts.

In all cases we can describe the value of fire prevention in terms of averting the direct, indirect, and intangible losses that occur from fire.

## The Disparity Between Costs and Risk

At the beginning of this chapter, we considered a hypothetical Robinson Crusoe and the value he might place on fire protection for his shelter. It was an elementary illustration of choice—about where people are willing to expend their money, or their personal time. We are all making decisions on how much of our resources to expend and where, and the amount we assign fire prevention is determined by the relative level of importance we place on it.

How, then, does the relationship between our perceived risk and our actual risk affect the economic benefit we assign fire protection? Using the Robinson Crusoe Example, it is simple. If we perceive the level of risk to be high, then the amount of brush clearing we are willing to perform would naturally be higher.

If, on the other hand, our perception of risk is lower, then we might ignore the brush clearing entirely and focus on hunting and fishing—taking our chances that a fire would never destroy our personal shelter.

How, then, does this play out in the real world? People are making decisions every day based on their perception of their risk, and the relative value of preventive effort. According to U.S. Fire Administration figures, older adults are 2.5 times more likely to die in fires than the overall population. As Americans age, their fire risk increases. There is a relationship between risk and gender or race, as well. Older men are at a substantially higher risk of fire death than women, and African Americans are at much greater risk of dying in fires than whites. The strongest correlations exist with income levels (low income correlates with a higher risk of fire).

What are the attitudes and behaviors we witness in these populations regarding fire prevention? The numbers speak for themselves: they continue to experience fires and fire losses out of proportion to their share of the general population. Despite being at higher risk, these populations do not appear to value prevention in relation to their risk. They act instead based on a commonly held belief: that fire will never happen to them.

So in raw economic terms, the relative value that individuals place on prevention has little to do with their actual risk—but rather depends on whether or not they perceive that risk.

There is a wealth of information on the perception of risk, and behaviors that might be risky but afford personal benefit anyway. References on Wikipedia highlight studies on the topic that serve to illustrate why some people perceive risk to be lower, and why they may be willing to engage in higher risk activity even though they know the risks involved.[5]

It is not hard to understand. A person at the lower end of the socioeconomic scale may be much more concerned with feeding her family than with creating a home fire escape plan and testing the smoke alarm. So in economic terms, the fertile ground for fire prevention professionals to plow lies in educating people about the real level of risk they face, and getting them to increase the amount of time and energy they are willing to expend on fire preventive measures.

# The Economic Value of Prevention

The raw economic value of prevention efforts only becomes clear after years of effort. If we can provide evidence that fire loss trends have been lowered because of our efforts (more on this topic in Chapter 12), we have compelling arguments for the quantitative value of those efforts. How many more fires would have occurred, and how much more of a dollar loss would there have been, if our programs had not been in place? How much indirect loss could we have expected to occur as well—with those figures converted into money saved?

However, it may be impossible ever to draw those kinds of conclusions about the value of prevention programs, because of the variables involved. A good economy might drive fire incident rates down on its own. Some of the older buildings in a community might have been burned down—or torn down—making way for new construction built to modern standards and demonstrably safer than the old styles of **balloon construction** we saw in the past. Balloon construction refers to

**balloon construction**
■ the type of construction made of framed wood—without fire stops to prevent a rapid spread of fire—that was evident in older housing stock

wood-framed buildings without fire stops to prevent a rapid spread of fire—a type of construction that was typical in older housing stock.

There are ways to gather evidence that prevention programs produce results. One is to identify other economic factors that may show the value of prevention efforts.

For example, another way to look at the value of fire prevention (and suppression) methods has been described by Phil Schaenman of TriData Corporation. He first introduced the concept of measuring the value of mitigation efforts by documenting the value of property saved—instead of the value lost—in a fire.

For instance, a fire controlled by fire sprinklers might have saved an entire structure from certain fire loss—and the value of that building (less the cost of repairing any damage from water) would be the economic value of that fixed fire protection system.

The same could be said for fire suppression efforts provided by an active fire department. A crew arriving in time to prevent fire spread could make a claim for the portion of the building they were able to save—rather than recording only the portion that was lost.

However, caution is needed in this approach. A fire in a wastebasket in the common area of a shopping mall could hardly be considered as placing the entire property at risk. Consequently, claims that a fire suppression effort (fixed or fire department) saved a multimillion-dollar structure that was never at serious risk would be spurious.

Regardless of the limitations involved, we can begin to describe the benefits of our prevention programs in terms of dollars—and there are other substantive ways of doing so.

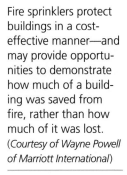

Fire sprinklers protect buildings in a cost-effective manner—and may provide opportunities to demonstrate how much of a building was saved from fire, rather than how much of it was lost. (*Courtesy of Wayne Powell of Marriott International*)

## TRADE-OFFS

A very real economic benefit to prevention programs may be found in the section of the fire code that allows for **alternate materials and methods**. This is an area where trade-offs allowed in the code may create a unique win-win situation for public safety, and for those in the construction industry.

Developers often wish to create narrower streets in developments to save money on paving. There are other interests sharing this wish that are more concerned with environmental issues (reducing nonpermeable surfaces helps water quality), but the economic issue for development is the cost of pavement, and the fact that wider roads mean fewer lots to develop and sell.

Many jurisdictions have found that allowing narrower roads, which still allow for fire department access, and reducing fire hydrant requirements make sense, when these concessions are made in exchange for fire sprinkler systems in the homes. In those cases, it often saves the developer money, while improving public safety for fire protection. As long as emergency crews can still arrive for medical responses, everyone comes out ahead. And the proposition that fire prevention efforts may provide a direct cost/benefit to communities is demonstrated.

Policies and practices that allow for this type of prevention advocacy pay off—in more ways than one.

**alternate materials and methods**
■ that portion of the code that allows local officials to accept different ways of achieving the same overall level of safety provided by the code

## INCENTIVES

Another way in which prevention can be affected by economic factors is through insurance. TriData Corporation has reported that Japanese and some European nations view the relationship between insurance and fire losses differently than the United States. In those cases, the building and contents may not be insured for their full value in case of a fire loss. In taking this approach, these nations have built in a sensitivity to fire safety because residents and business people know that they will suffer an economic loss if a fire should occur.

## Summary

There are a variety of ways in which economics influences fires and fire prevention programs. Some are measured directly, such as property losses attributed to fires. Others are measured indirectly, in terms of the ancillary losses that occur in relation to that fire—such as the loss of business to others while a fire-damaged company is closed. Other factors, like the loss of loved ones in a fire, cannot be measured economically, but have great intangible impact on a community all the same.

The cost of fire in the United States is measured by the dollars lost—and also by the money required to provide protection against fire. This aspect of defining the scope of the fire problem can help us demonstrate the overall impact that fire has on our economy—and the need to prevent fires wherever possible.

There are some economic values to fire prevention that are difficult to document but exist nonetheless. More is said in Chapter 12 about evaluating fire prevention programs, but it is possible to demonstrate the costs and benefits of fire prevention efforts in terms of tangible dollars. And there is a relationship between economic incentives and fire prevention, where some nations have linked fire insurance rates to fire losses in such a fashion that people have a built-in motivator to be fire safe.

The bottom line is: fire costs—and prevention is a cost-effective way to deal with it.

## Case Study

The City of Hatcher is considering a local ordinance requiring residential fire sprinklers. The local branch of the home builders and the real estate brokers associations have banded together to fight the prospect of the ordinance. They argue that fire sprinkler systems are far too expensive, adding thousands of dollars to the cost of housing. The figures they use are $4 to $6 per square foot to include fire sprinklers. They also argue that local water purveyors charge significant fees for the larger water meters required for residential fire sprinkler systems, and that those charges add an average of $2000 to the price of a home—regardless of size.

They feel that smoke alarms provide adequate protection, and note that most fires (nationally) occur in older homes, where fire sprinklers are not required to be installed.

One of the City Council members has been funded heavily by the home builders and realtors for her reelection campaign. There are five other council members, including the mayor, who will ultimately decide the fate of the ordinance.

## Case Study Review Questions

1. How would you go about evaluating the costs and benefits of fire sprinklers and making a case for or against them?
2. What specific arguments would you make about the fact that fires occur usually in older buildings—and that fire sprinklers will not help their overall safety?
3. How could you obtain accurate costs about installation costs for fire sprinklers?
4. Explain how you would handle the education process for council members on the topic of fire prevention, and the approach you might take given the fact that you are not a

contributor to political campaigns in the city in which you work.

5. Describe the costs and benefits of residential fire sprinklers, compared to the long-term cost of fire protection for a paid or volunteer fire department. Be prepared to make an argument in favor of one or the other in class.

6. How do economic factors affect fire rates and fire prevention efforts?

7. What is the difference between direct and indirect fire loss?

8. How do (and how should) we describe the cost of fire in the United States?

9. Explain the disparity between the high costs that fires inflict, versus the perception that most people have that their own risk of fire is low.

10. How can economic policies actually benefit fire prevention efforts?

## Endnotes

1. Yoder, J. (n.d.). Retrieved from http://www. forestencyclopedia.net/p/p805.

2. Wharton, Fred. (n.d.). "Insurance Fraud Prominent in Today's Economy." Retrieved from http://www.faia.com/core/ contentmanager/uploads/ Education%20Department/PDFs/Misc/ Insurance%20Fraud%20is% 20Prominent%20in%20Today.pdf.

3. Karter, Michael J., Jr. (August 2009). *Fire Loss in the United States During 2008*. Quincy, MA: NFPA.

4. Fire Loss Clock. (n.d.). Retrieved from http://www.nfpa.org/assets/files/PDF/ firelossclock.pdf.

5. N.A. Retrieved from http://en.wikipedia.org/ wiki/Perceived_risk.

# 11

# Governmental and Departmental Influences on Fire Prevention

## OBJECTIVES

After reading this chapter you should be able to:

- Describe the difference between private-sector and governmental influences on fire prevention.
- Describe the interrelationships among federal, state, local, and fire-department influences on fire prevention programs.
- Describe how local authority for fire prevention programs flows from national and state levels.
- Identify some key federal agencies that have influence on fire prevention programs.
- Identify state roles in fire protection and prevention.
- Describe how fire department structure and organizational culture influence fire prevention programs.
- Describe how fire prevention fits into community efforts and the importance of leadership for fire prevention programs.

Not all fire prevention programs are managed by the government. In fact, some of the most influential national "players" on the fire prevention scene are in the private sector.

**Private-sector influences** are those that do not have regulatory authority, but still influence fire prevention efforts in a variety of ways: by providing voluntary standards of performance or installation for fire protection features; by providing fire safety materials; and by testing products for safety.

private-sector
influences
■ lack regulatory
authority, but still
influence fire prevention
efforts in a variety of
ways

The National Fire Protection Association (NFPA) was formed in 1896, primarily by insurance interests, to standardize the rules for installing, testing, and maintaining fire sprinkler systems. Now this private nonprofit organization is one of the strongest influences on fire prevention efforts, in large part because of the scope of the code and standards development work conducted under its auspices. Many thousands of volunteers work on well over 200 voluntary safety standards that affect the fire prevention world by providing prescriptive installation and operational guidance for fire sprinklers, alarms, electrical systems, and a host of other construction-related features.

The NFPA also produces standards for fire protective clothing and equipment, including fire-engine and fire-truck specifications. They produce professional qualification standards that are based on job performance requirements for a variety of jobs within the fire service, including fire inspector, investigator, plan reviewer, and public educator. They also provide a host of educational materials for training fire prevention personnel and for educating the public about the dangers of fire and how to prevent fires. The "Sparky the Dog" character developed by the NFPA is as widely recognized as "Smokey Bear" for its relationship to fire prevention. The scope of NFPA's influence on fire prevention efforts is quite large.

The International Code Council (ICC) is another private-sector organization (nonprofit) that has tremendous influence on fire prevention programs. Primarily through the development of building, fire, plumbing, and mechanical codes, the ICC promulgates the building and fire codes that are adopted in most jurisdictions across the nation. Thus, it has a significant level of influence on fire prevention efforts, what might be called a defining role because it is responsible for the fire code regulations used in most places throughout the United States.

The insurance industry as a whole wields tremendous influence on fire prevention efforts in the United States as well. By determining the cost of fire insurance and how it is provided, the insurance industry plays a critical role in how the nation views fire safety. Their role in research about how fires begin, and in developing both regulations to govern fire safety and educational products that are designed to enhance it, is a significant part of how the United States prepares for fire

and fire prevention. Some insurance entities, such as FM Global, provide not only insurance, but engineering and loss control expertise for their clients to make them more fire safe, thereby reducing the amount of money needed for insurance premiums and fire loss payouts as part of an insurance settlement when fires occur.

It is in this way that insurance companies can have a direct impact on fire safety. Many also support fire prevention programs at a national or local level for similar reasons. A safer community or business will be less likely to need insurance-provided reconstruction funds.

The more cynical among the fire prevention community would argue that insurance pools actually benefit from fire losses because those losses increase the funding necessary for the pool—funding that must be invested and managed, providing profits for the insurance company. But this is speculation, mentioned only as another possible role that insurance plays in fire protection and prevention in our society.

There are other important private-sector organizations that influence fire prevention efforts. Underwriters Laboratories is one of the largest safety testing laboratories in the world. The UL symbol serves as an imprimatur of legitimate technological testing and performance for local jurisdictions across the nation. When a product has the UL symbol attached, local fire marshals everywhere recognize that the product that bears it has undergone rigorous safety and performance testing to ensure that it will not be a hazard to the public. Testing laboratories such as UL make a much-needed contribution to the overall fire safety of our nation by verifying that products and fire protection systems work as intended. There are other accredited testing laboratories, such as Intertek; so local fire prevention professionals should keep in mind that there may be more than one method of providing assurances that a product will perform reliably and safely.

All of these private-sector organizations have one simple thing in common: they are voluntary. Their usefulness comes from providing financial incentives such as insurance, or guidance for product liability, that can influence fire safety. But most of them function to one degree or another because their roles have been formally or informally accepted as legitimate by government agencies.

The codes and standards developed by nonprofit organizations must be adopted and used by someone. Our society has assigned the role of regulation to our government. We look to government for rules that will govern our behaviors without infringing on our freedoms. Consequently, government plays a powerful role in fire prevention efforts.

# Private Fire Departments

There are private fire departments throughout the United States, and as such they have a direct influence on fire prevention efforts in their own jurisdictions and possibly beyond, if they are actively involved in fire prevention issues outside their jurisdiction. For example, until fairly recently, the City of Scottsdale, Arizona, utilized a private company to provide fire protection for their community. They had one of the most aggressive fire sprinkler requirements for residential property in the nation. As a result, their long experience with fire sprinklers became a case study of some note, affecting fire sprinkler advocacy for many different jurisdictions and national organizations. There are also private fire departments (sometimes called fire brigades) that provide protection and prevention programs for large companies and universities in a variety of locations throughout the nation.

These private fire departments follow corporate rules rather than government regulations. But their efforts may be just as effective, and in some cases perhaps even more so. In addition, many private companies, including major hotel chains, provide their own fire prevention inspection services. Some, such as Marriott International, are among the most aggressive in identifying and correcting fire hazards. Marriott has long been involved in fire sprinkler issues inside and outside their company. They not only installed fire sprinklers before they were required by fire codes, they also help organize fire protection studies of fire sprinklers with the intention of improving the industry.

# Governmental Influences

Among its other functions, government exists to provide the rules under which our society operates, giving government a unique position when it comes to fire protection and fire prevention.

Government may provide services just as the private sector does—but no other entity governs our behavior with the force of law. **Governmental influences** on fire prevention are those that have some direct regulatory impact or the force of legal authority governing fire prevention efforts.

**governmental influences**
- regulatory power that governments exert over fire prevention efforts

The various levels of government can all influence the scope and nature of fire prevention programs. This chapter deals with four levels: federal, state, local, and departmental.

## FEDERAL-LEVEL INFLUENCES

The United States is a democratic republic. This means that the head of state and other political leaders are not chosen by heredity, but by a democratic election process. The form of the government is a practical republic, which means it derives its power from the people. The topic of government in general is a matter for other texts, but in this context it is important to note that the federal influences discussed are derived from a national view of government that is a constant balancing effort to provide rules of behavior for the benefit and safety of society, without infringing on personal freedoms.

Generally speaking, the authority to govern (by setting fire codes, for example) comes from the federal government to states and is passed on to local authorities having jurisdiction. That could be cities, townships, or special service districts formed specifically for special services such as fire protection.

This federal level of authority and the ensuing balancing act can be illustrated by the landmark report *America Burning*, issued in 1973 by the National Commission on Fire Prevention and Control. The report called for a strong national focus on the nation's fire problem. It also emphasized the need for primary reliance on state and local government in matters of fire safety.

> Emphatically, what is not needed is a Federal bureaucracy assuming responsibilities that should be retained by State and local jurisdictions. Fire prevention, fire suppression, and public education on fire safety should remain primarily responsibilities of local governments, where familiarity exists with local conditions and the people being served. Communities have already invested heavily in manpower and equipment for fire protection, in recognition that it is a local responsibility. Likewise, regulatory responsibilities for fire prevention and code enforcement should remain at State and local levels. Codes and regulations must respond to changes in the built environment, and past experience illustrates that State and local governments are likely to be more dynamic and responsive to changing needs for different jurisdictions than a single Federal agency.

Ultimately, the U.S. Fire Administration and the National Fire Academy grew out of that report, and the efforts of those behind it. But they clearly recognized that the role of federal agencies involved in the issue was not to wield supreme power over fire protection issues, but rather to provide a focal point for efforts and an ability to coordinate at a national level—as well as services that could not be provided at the state or local level.

A number of federal agencies have had an impact on fire protection and prevention for many years, beginning even before the formation of the U.S. Fire Administration. Some of these agencies are discussed next, along with their influence fire prevention.

## FEDERAL AGENCIES THAT INFLUENCE FIRE PREVENTION

### Internal Revenue Service

The resources of the IRS generally are not available to state or local investigators. However, in arson cases, the IRS has helped local agencies in the investigation of fraud. The IRS commonly works with other federal agencies more directly responsible for public safety, such as the Federal Bureau of Investigation or the Bureau of Alcohol, Tobacco, Firearms and Explosives.

### Federal Bureau of Investigation

The FBI has actively participated in fire investigation when criminal behavior (i.e., arson) is suspected and it falls within their jurisdiction. They have partnered with arson task forces at local levels all across the nation and can be actively involved in investigations of large-loss fires with aspects that cross state lines.

The FBI also operates the Uniform Crime Reporting system, established in 1930 to give police chiefs and others an index of U.S. crime trends. The Anti-Arson Act of 1982 made arson a "part one" crime under the FBI guidelines, and arson has been part of the principal crime reporting system ever since. The FBI also offers courses related to arson control at the FBI Academy in Quantico, Virginia.

## Bureau of Alcohol, Tobacco, Firearms and Explosives

The BATF is also concerned with arson, with a particular focus on explosives. The Omnibus Control Act of 1970 gave BATF responsibility for interstate control of explosives, so all commercial users must operate under their regulations for storage and interstate shipments.

The BATF also has National Arson Response teams that respond to major incidents of arson and bombings anywhere in the United States. They also work closely with local investigators on a regular basis when accelerants or explosives are suspected. BATF also supports the acquisition and training of accelerant-detecting dogs for local and regional fire investigation programs. In addition, the Church Arson Prevention Act of 1996 (which came about because of a series of arson in largely African American churches in the southern states) provided more authority for the BATF and the FBI to assume jurisdiction in any fire in a church to ensure that arson was properly identified or ruled out. In such cases, BATF routinely works with other agencies, including local investigators.

## Department of Defense

The Department of Defense provides fire safety research and regulations for military property all over the world. Their experience with fire prevention has been studied and recognized in reports from TriData Corporation and the results of their efforts have provided examples for others to follow. According to Phil Schaenman, president of TriData, the military has produced positive results with aggressive public education and code enforcement activities. More recently, studies at U.S. Air Force facilities using kitchen cooking safety appliances have provided a graphic example of how technology can be used to control, and even eliminate, kitchen cooking fires.[1]

## Department of Agriculture

The Department of Agriculture oversees the Forest Service and the fire problems and prevention measures related to forests. The Forest Service operates the largest firefighting force in the nation: the thousands of personnel engaged in forest firefighting every year. The Forest Service is also known for working with the Ad Council (www.adcouncil.org) on developing a prominent public fire safety campaign that established Smoky Bear as one of the most recognizable fire prevention characters in the world. Smokey Bear was developed in 1944 and has reached millions of homes over the years since it was begun as a campaign to educate the public about the dangers of wildfires—and more important, how they can be prevented.

The Smokey Bear safety campaign can be researched on the Internet, where a variety of public education messages and materials are still available to teach important lessons about preventing wildfires.

Smokey Bear is often mistakenly referred to as Smokey *the* Bear from the 1952 song of that name. Accounts say that songwriters Steve Nelson and Jack Rollins added "the" to fit the rhythm of the melody. The Smokey Bear campaign is widely viewed as one of the most effective public safety campaigns in existence. (*Smokey Bear image used with permission of the USDA Forest Service*)

The Department of Transportation influences prevention efforts in one way by regulating the transport of hazardous materials and their required identification methods.

## Department of Transportation

The Department of Transportation (www.dot.gov) plays a critical role in fire prevention and protection through its regulation of the transport of hazardous materials. The rules governing transport and notification of emergency responders about the types and dangers of hazardous cargo have been an important foundation of fire protection efforts for many years. Firefighters everywhere count on the safety placards to help identify the risks they may encounter when responding to am emergency incident involving commercial hauling equipment. The placard system helps them identify, control, and mitigate further damage from a wide range of hazardous materials when they are involved in a vehicle collision or crash.

## Department of Health and Human Services

The Department of Health and Human Services is responsible for funding the care of many patients in nursing homes. Because the agency is responsible for general patient care, they also require inspections to ensure that these facilities are reasonably safe. Since the 1970s, hospitals and nursing homes have been required to comply with NFPA 101, the Life Safety Code.

These inspections are often conducted by state personnel trained to do so under NFPA 101. Sometimes the same occupancies are also regulated by the locally adopted fire code—these days, most commonly, the International Fire Code (IFC). Many of the provisions in NFPA 101 and the IFC are the same or similar, but where conflict occurs, local caregivers may be confused about jurisdiction and competing orders for safety.

This overlap in jurisdiction and rules governing safety in medical facilities among HHS, state governments (often a state fire marshal), and local fire marshals is a prime example of how federal, state, and local governments have the potential to get in each other's way. One effective way to manage these potential conflicts has been demonstrated by numerous jurisdictions across the nation, who coordinate their fire safety inspections and requirements between the two source documents (NFPA 101 and the IFC) to avoid confusion of care providers, and to ensure that safety regulations are met at every level.

## National Transportation Safety Board

When an airliner crashes or a train derails, the NTSB is called in to investigate. They are responsible and authorized by federal law to investigate significant transportation incidents to determine the cause. Often the findings from the investigations provide the basis for new safety features or regulations designed to prevent similar occurrences.

At their Web site, the NTSB also provides a focus on safety by maintaining a list of improvements they recommend be adopted for highway and marine transport, aviation, and pipelines to reduce risks from collisions, crashes, and hazardous material spills and fires.

Though not focused so much on fire, the NTSB plays an important safety role in many of the types of emergency calls responded to by fire departments across the nation.

The National Transportation Safety Board investigates large incidents involving marine, rail, or air travel to determine causes and help provide guidance for prevention or mitigation measures.

## Consumer Product Safety Commission

The CPSC was created by Congress in 1972 to protect the general public from what would be considered unreasonable risks associated with some consumer products. The CPSC board is composed of political appointees, and thus will sometimes be influenced by the political will of the day and other national laws. For example, although cigarettes are a consumer product, the CPSC was forbidden to deal with cigarette safety standards for many years when fire prevention advocates were attempting to change the design of cigarettes to be less prone to start fires.

The CPSC develops voluntary and sometimes mandatory safety standards, occasionally banning some products and ordering them to be recalled. CPSC maintains a hotline for consumers to learn about unsafe products, and also maintains a staff of field representatives who conduct their own product safety investigations. These field representatives often work with fire department investigators to study how some products fail and start fires. The agency also operates the National Electronic Injury Surveillance System, with about 100 hospitals that have 24-hour emergency rooms. The system collects data on product-related injuries that ended up being treated in the ER.

CPSC pays particular attention to products that may cause fires, as well as failures in products that protect against them.

For example, in 2001 the CPSC entered into an agreement with the Central Sprinkler Company to recall many thousands of fire sprinkler heads that contained O-rings designed as a water seal, but that prevented some of the heads from operating. This voluntary recall ended up involving other sprinkler

The Consumer Product Safety Commission plays an important role in fire safety through voluntary or mandatory recall programs of defective products—including defective fire sprinklers. (*Courtesy of Rob Neale*)

manufacturers as well, and it demonstrated the role that CPSC could play in the fire protection arena.[2]

For years, the CPSC has also been involved in creating flammability standards for children's sleepwear, which has been controversial because carcinogens have been found in the fire retardants used in some clothing fabric. The CPSC enforces flammability standards for mattresses, and in the past the agency has been an important force in regulating other products such as electric blankets or coffee makers that were known to cause fires.

## National Institute of Standards and Technology

NIST is the home of the national Building and Fire Research Laboratory (BFRL). The center performs research on materials, fire detection methods, fire propagation and even human aspects in fire problems.

One of the strategic goals of the BFRL is to examine innovative fire protection methods. According to the laboratory's Web site:

> Fires continue to kill more people per capita in the U.S. (by as much as a factor of two) than in most other developed nations. Fire losses from systemic causes are preventable.
>
> Significant damage from wildland-urban interface (WUI) fires is on the rise in the U.S. and there have been two major WUI fire loss events in the last five years. The 2003 Cedar fire in California cost $2 billion in insured losses and destroyed

3,600 homes, while the October 2007 southern California fires displaced residents of over 300,000 homes. Overall, the trends suggest that the severity of the U.S. fire problem is growing.

There is an incomplete understanding of fire behavior, which hinders the development of innovative fire protection. Current prescriptive fire standards and codes stifle innovation in fire safety systems, technologies, and building design. To ensure fire safety in a cost-effective manner and to reduce fire losses, it is essential that adequate science-based tools are developed to enable the implementation of the next generation of standards, codes, and technologies that address the U.S. fire problem. Measurement science is lacking to reduce the risk of fire spread in buildings, to reduce fire spread in WUI communities, to ensure effective and safe use of emerging fire service technologies, and to derive lessons from fire investigations.

Addressing these measurement science needs is essential, if fire losses are to be reduced and the resilience of buildings and communities (people and property) are to be increased.

The role of NIST and the BFRL is very important in the fire protection and prevention community. The level of science they are able to apply with regard to fire safety raises the bar for everyone associated with the field.

For example, NIST has been closely involved in studies on the effectiveness of different types of smoke alarms, most notably ionization and photoelectric technologies. Currently, there is a great deal of controversy among fire prevention professionals over which technology functions best overall, mainly because some studies have demonstrated that ionization alarms have tended to operate slowly in smoldering fires. NIST is able to guide studies, and conduct their own, with a relatively neutral and unbiased view because of their role as a government agency.

## The U.S. Fire Administration

The Department of Homeland Security houses the Federal Emergency Management Agency. FEMA houses the U.S. Fire Administration (USFA), which in turn houses the National Fire Academy (NFA).

The U.S. Fire Administration was actually formed in 1974 under the Department of Commerce as the National Fire Prevention and Control Administration. Fire prevention was one of its principal responsibilities, which included the development of the National Fire Incident Reporting System (NFIRS) and the data center at USFA for its analysis. The USFA also manages the Learning Resource Center (LRC) on the grounds of the National Emergency Training Center, where the Emergency Management Institute and the National Fire Academy are located (in Emmitsburg, MD). The LRC is a tremendous resource for materials on fire prevention, including research papers and an online version for searches that can yield many free resources for local prevention efforts.

The Web page for the USFA is full of free publications, studies, and materials for fire prevention efforts, including specialized materials such as those produced for juvenile firesetting, or the annual fire safety campaign materials on specific topics such as smoking safety. The USFA also produces regular messages to go along with news topics, so that local prevention professionals have access to ready-made educational efforts that can be used whenever a particular fire cause occurs locally. For example, a fatal fire may occur because of a space heater located too close to combustibles. The media center at the USFA has accompanying messages designed to be used in conjunction with the media coverage of that tragedy, turning it into an educational opportunity to prevent future occurrences.

The U.S. Fire Administration also hosts a biannual conference, called PARADE, for state fire marshals and metro-sized community fire marshals. PARADE stands for Prevention, Advocacy, Resource And Data Exchange. Smaller jurisdictions are represented by fire marshals selected from the International Fire Marshals Association regions throughout the nation. The conference and accompanying conglomeration of fire prevention professionals throughout the nation provides a focal point for professional development, networking, and coordinated fire prevention activity in the United States.

The location of the U.S. Fire Administration within the Federal Emergency Management Agency is important because FEMA also has a tremendous influence on fire prevention efforts independent of its role as the parent organization for the USFA. FEMA currently houses the Assistance to Firefighters Grant (AFG) program, which provides funding for fire protection programs nationwide. Set within the overall AFG grant program are some grants specifically designated for fire prevention programs. The Fire Prevention and Safety grants are allocated each year based on a peer review process that screens thousands of requests from local, state, and national fire prevention organizations. That includes individual fire departments, and nonprofit organizations with an active interest and role in fire prevention. These grants provide thousands of opportunities for solid financial assistance for fire prevention efforts nationwide.

FEMA was formed in 1979 as a Cabinet-level government agency to provide a single official and accountable source of federal emergency preparedness, mitigation, and response activity for all types of hazards that could be categorized as disasters. In 2003 FEMA and 21 other federal government agencies were consolidated under the Department of Homeland Security to provide a coordinated focus for every government agency that could have a role in preparing for, mitigating, or responding to acts of terrorism.

## STATE INFLUENCES ON FIRE PREVENTION

**state influences**
■ provide leadership on fire prevention programs outside the code arena

**State influences** on prevention programs are significant because of the role of states in adopting codes and standards. Many states provide leadership of fire prevention programs outside the code arena, such as South Carolina, which has documented significant reductions in fire deaths over the past two decades.

As mentioned in Chapter 3, the authority to adopt codes and standards into law flows from the federal level, to the state and on to local jurisdictions. Thus, states can exert considerable influence on the fire prevention efforts of any local jurisdiction. But the level of authority retained in each state varies widely. Usually fire prevention efforts, and especially fire code enforcement, are found in the offices of a state fire marshal. But not all states establish a state fire marshal position—and even those that do vary widely in their approach to handling fire prevention.

The National Association of State Fire Marshals (NASFM) describes the situation on their Web page.[3] A state fire marshal may have various responsibilities including fire code adoption and enforcement, or fire investigation responsibilities.

Many are responsible for collection of fire loss data through the National Fire Incident Reporting System (NFIRS). Some (such as in South Carolina) are very active in public education activities designed to reduce fire incidents. Some are responsible for firefighter training and certification, and some (as in Oregon) are responsible for mobilizing large firefighting forces for wildland fires.

Some are appointed by governors, and others by high-level state managers. Some are housed within the state police, and others in separate agencies or independently. The variety of jobs and tasks that each person known as the "state fire marshal" varies considerably across the nation.

Because the job titles, functions, and duties vary so widely, it is difficult to describe the full scope of how states can influence fire prevention. Aside from code adoption, every state has a legislative body that establishes law. If nothing else, the state, and through it the state fire marshal (or an equivalent official) may wield a great deal of influence over the legislative process.

As an example, many states are currently embroiled in controversy over requirements in the International Residential (construction) Code that mandate the inclusion of fire sprinkler systems in all one- and two-family dwellings. The largest and most influential organization opposed to these requirements is the National Association of Home Builders (NAHB). The debate did not end with the adoption of the requirement in the model (IRC) code. It has now moved on to each state in the code adoption process. In many states legislation has been introduced, and in some cases passed, that would omit those requirements.

In this debate the state fire marshal may or may not have influence over the outcome of the fire sprinkler requirement. But the state can enact law that directly determines the amount of fire protection required in new construction.

States influence fire prevention efforts in a variety of ways, but perhaps most significantly through the legislative process, where safety laws may be enacted—or fought.

"Novelty" lighters have been banned in several states because they are sometimes mistaken for toys by small children who may use them inappropriately, causing tragic fire losses. (*Courtesy of Portland Fire & Rescue, OR*)

There are numerous other examples of how the state can influence fire prevention efforts through legislative action. Many states have laws regarding smoke alarms that go well beyond new construction and mandate their use in every home. Often these requirements are not well enforced because of a lack of resources and the difficulty of gaining entrance into private property. Most often these types of requirements are enforced at the time of sale of a private property.

Some states have been dealing with laws that would prohibit the sale of "novelty" lighters. These are lighters that are made to look like a toy of some type. Children have sometimes mistaken them for toys and started fires as a result of innocent play. The U.S. Fire Administration maintains a Web site devoted to tracking how many states have passed this type of legislation.

In a few states, such as Florida, legislation has been passed regarding the level of training fire prevention professionals must obtain. Florida requires that everyone conducting a fire inspection be certified as a fire inspector in that state. The Florida Fire Marshals and Inspectors Association provides ongoing training for its members to meet that requirement, and thus has secured a steady source of income to support its fire prevention efforts while improving the professionalism of its members.

To one degree or another, states (usually through their state fire marshal or an equivalent) provide a number of resources to support fire prevention efforts at a local level. The state fire marshal's office in South Carolina, for example, provides a number of direct public safety education programs dealing with high-risk audiences. As explained on their Web site[4]:

> Throughout the 1980's, South Carolina historically ranked among the top three states in the nation having the highest fire related deaths, often ranking in first place. Many factors attributed to this ranking: the vast rural areas with volunteer fire services, low educational levels, and high poverty levels. In 1988, under the leadership of Governor Carroll Campbell, $50,000 was appropriated to combat the seriousness of fire fatalities, beginning the first statewide initiatives of public fire education. It was determined that the need for statewide coordination is critical in that approximately 45 percent of South Carolina's population lives in rural areas served mostly by all-volunteer fire departments. Without full-time staff dedicated to fire prevention, these citizens are largely underserved. Departments with full-time public fire safety educators are an exception.
>
> Statistics point out that the public fire education has made a significant impact on today's fire death experiences. In 1988, 168 fire fatalities were reported, while in 2002, 92 fire fatalities were reported. More important is the break in the upward fire death trend experienced in the early 1980's. There has been an overall downward trend in fire fatalities since 1988, when the first statewide fire education effort began.

In 2008 South Carolina reported 74 civilian fire fatalities, a noteworthy downward trend in fire deaths for their state—once considered one of the high-risk areas for fire losses.

Other states provide varying levels of support for fire prevention efforts, including most that collect and analyze fire loss data through the National Fire Incident Reporting System. The overall point is that most states may have significant influence over fire prevention efforts, though each may have a unique approach.

## LOCAL AGENCIES' INFLUENCE ON FIRE PREVENTION

The authority to adopt and enforce a fire code flows from the federal to the state level and on to the local. Several kinds of local repository for that authority exist. Sometimes authority for fire codes resides in a county government. In Washington state, for example, the position of fire marshal and the authority to enforce the fire code is derived from the state and given first to counties. It is beyond that level of government that municipalities may exercise their option of providing their own fire code enforcement services. To varying degrees, some counties take on that responsibility and establish a formal fire marshal's office. Others ignore the state law, at least in part, and relegate that role to another government agency or do not provide that service at all.

Other states provide examples of how a county can influence or even lead fire prevention efforts. In Nassau County, New York, the office of the fire marshal provides a centralized area of prevention programs for smaller jurisdictions throughout the county. Their Web site expands on the variety of services they provide, including county-wide training and prevention services. The office of the fire marshal conducts code enforcement inspections throughout the county, investigates fires on request by local fire departments, provides hazardous materials response, and even dispatches services for the county.

When a jurisdiction reaches a certain size, it generally incorporates into some form of municipality and quite often reserves the fire suppression and fire prevention duties for itself. Some have said that all fires occur locally, and so all fire prevention efforts should be local. It is fair to conclude that local agencies may ultimately wield the largest influence of all on fire prevention efforts, because they are closest to the point of contact where prevention efforts will do the most good: the individual citizens who start or are harmed by fire.

# Fire-Department Influences on Fire Prevention

**fire-department influences**
■ a significant factor because of the level of importance they place on prevention programs

**Fire-department influences** on fire prevention may be the most significant factor because of the level of importance fire departments give to prevention programs. With more than 30,000 fire departments in the nation (according to U.S. Fire Administration estimates), a local fire department can make or break fire prevention programs just by deciding to initiate them, or to cut them because of budget constraints or a lack of interest.

Chapter 9 covers the topic of organizational culture and its role in determining the level of importance given to prevention by fire department members and leadership. This is the informal—and sometimes the most powerful—influence that an individual department may have over fire prevention efforts. An organization that does not value prevention will not support it to any great extent.

But the formal structure of the department, and where fire prevention programs fit within that structure, is the official notice of how important fire prevention *should* be. It usually begins with a mission statement found in whatever kind of planning document (strategic or other) an individual fire department uses.

Part of the mission statement of the Phoenix Fire Department in Arizona looks like this[5]:

> The Phoenix Fire Department is committed to providing the highest level of public safety services for our community. We protect lives and property through fire suppression, emergency medical and transportation services, disaster management, fire prevention and public education.

Note that fire prevention and public education are both included in the mission statement, indicating their high value within the formal organizational structure. So a mission statement is often the first step in securing and identifying the organizational role of fire prevention in any fire department. Certainly, not all fire departments provide prevention services in any form. Most U.S. fire departments are staffed with volunteers. Even in such cases, a mission statement might reflect the organizational goal of preventing fires from occurring in the first place. But evidence for delivery on that stated goal will be found somewhere else. In short, a mission statement is important but not the only—or even the truest—indicator of the importance that prevention receives within a fire department or public safety agency.

The next way a formal structure can reveal the importance of fire prevention is in an organizational chart. The accompanying figure shows two hypothetical examples of a fire department organizational structure.

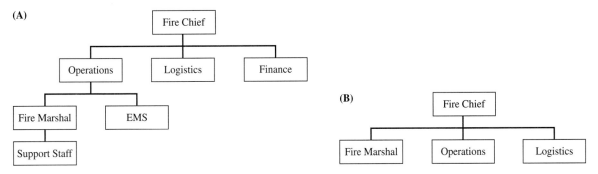

**(A and B)** The two organizational charts shown here provide clues to the level of importance fire prevention efforts receive from each of these fire departments.

Each structure provides a clue to the level of importance fire prevention receives in the organization. It is relatively simple to deduce which department values fire prevention more—at least on the surface. The closer to the head of the organization that prevention efforts are placed, the higher level of support they will usually receive—again, at least on the surface.

The organization mission and structure are not automatically indicators of the true level of importance given to prevention. One ultimate measure is the percentage of the budget devoted to prevention. However, there are many ways to support prevention without direct budget dollars. For years, the Plano, Texas, Fire Department has supported their public education efforts by using on-duty firefighters to perform a variety of public fire safety programs. Dallas, Texas, also uses on-duty firefighters to perform public safety education activities, including home checks and installation of smoke alarms.

Ultimately, the formal structure may indicate how important fire prevention *should* be for any organization, but the reality probably lies somewhere below the surface view.

## LAW ENFORCEMENT STATUS

Aside from the status that prevention receives, there are other ways that fire departments influence fire prevention efforts. One way is the level of law enforcement powers the department's members receive. This is not an automatic discriminator, because there are other ways to prevent fires. But it is difficult, if not impossible, to administer the fire code provisions if the local jurisdiction has not legally adopted the fire code and vested itself (through local government) with the police powers necessary to actually enforce it.

Another area where law enforcement status may play an important role is for fire investigations. If a local jurisdiction will actually investigate fires, sooner or later they will run across a case of arson—a deliberately caused fire. When they do, either they will be forced to turn the investigation entirely over to a police agency, or they will have to secure their place in the hierarchy of criminal investigation, arrest, and prosecution. Assigning police powers to fire investigators inside a fire agency can be a mixed blessing. Police powers can improve an organization's ability to actively contribute to investigations and prosecutions of arson, thereby deterring arson in their community. But police powers also come with a price tag for increased training and the potential (at least) for dilution of investigation resources.

More is said about law enforcement status relative to prevention staffing in Chapter 13.

Ultimately, the level of importance fire prevention efforts have within their organization can be determined by the quality of its leadership. More will be said about that issue as well in Chapter 13.

## Summary

Both governmental agencies and private-sector organizations can heavily influence fire prevention efforts. But the role of enforcement is usually reserved for government, and that distinguishes government from private-sector, mostly voluntary efforts. The authority to adopt and enforce fire codes, for example, flows from the federal to the state and ultimately to the local level. And throughout that continuum there is an interrelationship between agencies where opportunities to cooperate on fire prevention efforts abound.

There are numerous federal governmental agencies that influence fire prevention efforts, sometimes in surprising ways. An example includes the Internal Revenue Service, which does not generally help local jurisdictions but has been a valuable partner on occasion for criminal fire investigations. Each state may also influence fire prevention efforts in unique ways. Some offer a wide range of prevention services, including training, data collection and analysis, inspections, and investigation services and public education. Others

offer minimal services, but may still play an important role in the legislative process where codes are often adopted, or independent laws regarding safety (e.g., retroactive smoke alarm laws) are created.

Ultimately, whatever power flows through the federal and state level to the local level will reside in some form of fire safety agency that functions at that level. Thus, informal and formal organizational mission and structure have perhaps the strongest influence on fire prevention efforts.

It is important to note that at nearly every level of government, federal, state, or local, other agencies (outside the fire service) also influence fire prevention efforts. Some of them are listed in this chapter, and there are more that can be found by a simple search of the Internet. It is a truth often ignored that fire prevention efforts are most effective when they are carried out in collaboration with others. More is said about leadership and collaboration in Chapter 13—but both are critical to the success of fire prevention efforts.

## Case Study

The fictional Parish of Lauziere, Louisiana, lies on the Gulf Coast in a largely rural area with a population of about 30,000 spread out over a 290-square-mile area. The citizens are mostly of European descent, but members of a large Cajun community speak their own form of French and English. There is also an extremely poor African American segment of the population that lives inland in poor housing with few amenities.

The recent hurricanes and accompanying floods in the area have devastated homes, businesses, and the infrastructure of the community. Residents are recovering slowly, but the poor in particular are hampered by limited resources, relatively low educational levels, and generally a limited understanding of how government operates—let alone how it might help them recover.

The Federal Emergency Management Agency (FEMA) provided temporary shelters (mobile homes) that were later found to contain relatively high levels of formaldehyde, allegedly causing sickness in some and at the very least causing a major political problem for national, state, and local leaders. Replacement housing needed to be found for the short term, and long-term housing was going to be a major challenge because many families in the community lacked insurance.

Parish leaders recently established a panel to discuss the housing situation and recovery efforts, and established for the first time an interim position of fire marshal to help them assess the nature of the problems they would encounter as they tried to rebuild the community. The parish already has a part-time building official, who is experienced but lacks formal certification or training and tends to rely on his construction background rather than the code for making decisions. There are three parish fire departments, all of them predominantly volunteer organizations, with only two paid fire chiefs and three paid firefighters in the entire parish. They are not well paid, and they supplement their income with part-time jobs outside the fire service.

Recently, a fire in temporary housing (a tent city) killed one family and injured several others, including children of both Cajun and African American communities. The people in the community are angrily demanding that something be done soon to provide help. The local, state, and national media have developed an active interest in the story because of the FEMA controversy and the recent fire deaths.

## Case Study Review Questions

1. How would a better understanding of the role FEMA plays in disaster recovery benefit this community?
2. Describe the relationship between FEMA's role in disaster recovery and the organizations and agencies involved in developing building and fire codes as they apply to housing in the community of Lauziere.
3. Given the case study, what information about temporary housing and smoke alarms would benefit the community, and where could it be found?
4. Explain how the local fire departments might feel about the appointment of a fire marshal, and how a collaborative environment might be achieved.

5. Explain what resources the state might have to support a review of affordable housing standards for the Parish of Lauziere.
6. List some federal agencies that may have an impact on fire prevention efforts at the local level and what that impact is.
7. Describe states' roles in fire prevention and protection.
8. Describe the difference between the impact of private-sector organizations and that of governmental agencies on local fire prevention efforts.
9. Describe the role that Underwriters Laboratories plays in fire prevention efforts.
10. How does the Department of Agriculture affect fire protection and prevention efforts?

## Endnotes

1. N.A. (August 28, 2009). "Pioneering's Safe-T-element® recognized by united states air force as a solution that meets the U.S. Military's Fire Safety Code requirements." MarketWire. Retrieved from www.marketwire.com/press-release/Pioneering-Technology-Corp-TSX-VENTURE-PIO-893594.htm.

2. N.A. (2001/2003). "CPSC, Central Sprinkler Company announce voluntary recall to replace O-Ring fire sprinklers." Release # 01-201. Originally issued July 19, 2001; revised May 28, 2003. Retrieved from http://www.cpsc.gov/cpscpub/prerel/prhtml01/01201.html.

3. N.A. (n.d.). Who are state fire marshals. Retrieved from http://www.firemarshals.org/.

4. N.A. (n.d.). South Carolina Department of Labor, Licensing and Regulation. Office of State Fire Marshal. Retrieved from http://www.llr.state.sc.us/fmarshal.

5. N.A. (n.d.). *City of Phoenix: Official Web Site*. Retrieved from www.phoenix.gov/fire/.

# CHAPTER 12

# Evaluating Fire Prevention Programs

## KEY TERMS

Benchmarking *p. 219*

Formative evaluation *p. 220*

Government Accounting
Standards Board *p. 219*

Impact evaluation *p. 232*

Outcome evaluation *p. 232*

Performance measurement *p. 219*

Process evaluation *p. 232*

Workload, efficiency, and
effectiveness measures *p. 219*

## OBJECTIVES

After reading this chapter you should be able to:

- Describe how the process of evaluation is linked to proper planning.
- Identify workload, efficiency, and effectiveness measures, including resource changes, educational gain, risk reduction, and loss reduction.
- Describe the value of benchmarking.
- Describe the difference between performance measurement and evaluation for fire prevention and life safety education programs.

PEARSON
**myfirekit**

Today, local decision makers face extreme pressure to justify expenses for every type of government service. Concerned taxpayers want to know the outcome produced by their tax dollars and whether the funds are managed efficiently. This level of concern has spawned more than one tax-reduction initiative in many areas throughout North America. People want to see results for their tax dollars. "Results" usually is translated into: What are these tax dollars doing for me? What public good is accomplished, and how does it affect my life?

Measuring results is necessary because we have developed programs with some specific end goals in sight. The relationship between planning and evaluation is discussed in other chapters, but it is important to note that evaluation is actually a part of the planning process. It is dealt with separately here merely to allow the detailed discussion necessary for an understanding of the subject matter.

Against this backdrop, fire prevention programs are often difficult to evaluate, even though most people accept the concept that preventing an incident is cheaper than dealing with it after the fact. Within the fire service, prevention programs (in North America) always receive far less funding than emergency operations, and when cuts do occur, prevention programs are often the first to be cut. It should not be surprising that most of the money in local fire departments go toward emergency response. In fact, even the most avid fire prevention professionals must admit that their departments were formed principally to fight fires after they occur.

Practical politics is an important factor in these decisions. For example, when faced with a choice between cutting a fire station or cutting prevention programs, a local political leader (and other decision makers) will face far more criticism for cutting the fire station. What constituency will complain to elected officials that they are not getting fire inspections frequently enough? The business community generally will not complain about a lack of inspections—unless their business depends on it. Some of the logical stakeholders with an interest include fire protection contractors who have difficulty getting property owners to pay for annual inspection, maintenance, and testing of fire protection systems. Without a regular inspection program, there are no local "teeth" to enforce those requirements, so the fire protection contractors' business may languish as a result. But other than that, supporters of regular fire code compliance inspections may be hard to find.

The only notable difference for prevention programs and widespread business support is in the development community. If business developers must comply with building and fire codes, they usually want prompt and competent service, and so they will complain about development being held up because of a lack of resources devoted to new construction. For other areas, such as fire investigations and public education, stakeholders who are motivated to complain about a lack of resources are fewer and farther removed. Investigations may be mandated by law, but sometimes the law can be overlooked when resources just do not exist. Yet everyone in the community has an active interest in having professional help (paid or volunteer) available to respond in an emergency. So it is not hard to figure out why the funding priorities are the way they are.

It is not that prevention is considered unimportant; rather, the political constituency is lacking, and therefore it is usually left to the more passionate believers in the old prevention adage "an ounce of prevention is worth a pound of cure" to actively support prevention efforts no matter what. Without that public pressure, fire prevention efforts can fall behind during tight budget times. And to gain public trust and support, prevention efforts must demonstrate their worth.

Consequently, the pressure is great to establish efficient prevention programs that produce results. Many fire departments have found ways to combine their emergency response and prevention efforts as a way to leverage additional resources. Others have been creative about combining building and fire code compliance efforts—though this should be done with caution because of the complexities of the independent codes, even when they are correlated. More is said about the relationship between building and fire codes in Chapter 4. The point is that, although efficiency is an issue because of limited budgets, the pressure to demonstrate results will always be great.

There *are* performance measures available that can indicate whether prevention programs are producing desired results and doing so in an efficient manner. This chapter outlines several aspects of the evaluation process, including the difference between performance measurement and evaluation of specific prevention programs. These methodologies include the general performance measurement approaches advocated by the Government Accounting Standards Board and the American Association of Government Accountants. That includes issues such as benchmarking and workload, efficiency, and effectiveness measures. This chapter also covers the important issues of data collection, documenting changes in resources, and the fundamental measures for effectiveness of prevention programs, which include educational gain, risk reduction, and loss reduction. It also deals with *program-specific* measures often promoted by the public health community as they relate to evaluation of more specific prevention efforts. Those measures include formative, process, impact, and outcome evaluations.

# Relationship Between Planning and Evaluation

It is often said that good programs begin with the end in mind. By the same logic, good evaluations depend on what the program was designed to deal with in the first place. There is an inextricable link between good program planning (as discussed in Chapter 7) and the evaluation measures used to determine the results that same program has achieved. Evaluating a program for results it was not intended to produce is misleading at best and pointless for those wanting to demonstrate actual results of prevention efforts.

Those wishing to properly evaluate their prevention programs should understand this link. A simple example might look something like this: the desire of a prevention program is to reduce fire incidents in the community, and program developers envision evaluating it in that fashion. But the program plan failed to address the fact that a coloring book or board game on fire safety quite possibly will not appeal to a senior audience. Consequently, the results of the effort were probably questionable from the beginning.

Proper planning, including formative research to examine what is needed, leads to program design that is more likely to be effective. It will then be possible to evaluate the impacts or outcomes those programs produce.

# General Performance Measures

The **Government Accounting Standards Board** (GASB) has produced several evaluation measures that can be applied to fire department activities nationwide. Looking at a specific program in one geographic area is more direct because the variables for measuring results can be more easily quantified. But comparing them from different jurisdictions is more challenging.

GASB is a private nonprofit organization devoted to developing and promoting model performance measures that aid local decision makers in documenting the results of programs supported by tax dollars. They usually work with financial and/or performance auditors from local or state governmental agencies.

Problems arise when we try to gain an understanding of the overall performance of prevention programs in different parts of the nation. One such problem is the need for common terminology. GASB has stipulated that it is difficult to compare one jurisdiction with another, because the variables in different jurisdictions are hard to match. In order to understand why this is, we must first look at what most fire departments have in common.

National studies, most notably from the National Fire Protection Association, indicate that income levels are the most significant factor when comparing risk of fire losses. That has been generally true from one jurisdiction to another. However, the age of structures, other socioeconomic factors, the climate, and the size of the jurisdiction also are important comparative features. Finding two jurisdictions that match in all of these categories is extremely unlikely. However, there are ways of evaluating fire and life safety prevention efforts that work well and begin to give us some common language to describe the results of our efforts.

GASB has developed a series of measures that they categorize into service, effort, and accomplishment reporting. They have promoted some terms that are generic enough to be used from one jurisdiction to another. GASB tends to focus on overall performance, rather than the measurable results of a single prevention program. That is true because government accountants and auditors want to measure the performance of programs on a broader level. This type of evaluation is called **performance measurement**.

Generally speaking, the combination of all these performance indicators can be categorized as **workload, efficiency, and effectiveness measures**. Additionally, many jurisdictions are using **benchmarking** as a way to measure the progress and impact of their programs, and GASB and the Association of Government Accountants have promoted that way of looking at overall performance through their *service, efforts, and accomplishment* reporting. In this context, performance measurement includes benchmarking (comparing one program to another) using the specified terms.

*Workload* measures are those that document the amount of work conducted. These include measures such as the number of code enforcement inspections done per inspector or the number of presentations performed by public educators. *Efficiency* measures are those that demonstrate whether something is done quickly and at the lowest possible cost. These include the cost per inspection or the cost per public education presentation. *Effectiveness* measures examine the more "solid" results from the perspective of local decision makers when they are evaluating their prevention efforts. Effectiveness measures show the impacts and

**Government Accounting Standards Board**
■ a private nonprofit organization devoted to developing and promoting model performance measures that aid local decision makers in documenting the results of programs supported by tax dollars

**performance measurement**
■ documenting results programs produce in specific terms

**workload, efficiency, and effectiveness measures**
■ the specific terms government performance auditors usually use when describing the results of programs

**benchmarking**
■ the process of comparative analysis—one jurisdiction to another or one to itself

outcomes of those efforts and their relationship to their stated goals and objective. Effectiveness measures get at the heart of the question usually asked by concerned taxpayers: Why is this service in place and what is it providing us? In prevention terms, the question regarding results would be: Are we really preventing losses?

> *Benchmarking* is a term used to describe the comparative results of a program. Comparing fire incident rates in one jurisdiction to those in another or to the same jurisdiction's previous incident history would be examples of benchmarking.

Establishing goals and measuring them compared to where they have been or whether they have been achieved accomplishes a form of benchmarking to one's own history. Benchmarking is also used to compare one jurisdiction to another—to measure our results in terms of how our efforts stack up against those of other jurisdictions. Benchmarking is an attractive way to evaluate programs for many decision makers, because it allows them to see how local programs perform relative to others. We discuss benchmarking in more detail later in the chapter, but a cautionary note is in place here: persons using exclusively benchmarking as an evaluation tool should be cautioned that a lack of matching variables between some apparently comparable jurisdictions may make meaningful comparisons extremely difficult.

# Data Collection and Analysis

Efforts to evaluate the impact of life safety and fire prevention programs begin with the collection of data. It is very important for practitioners to understand that the process of evaluation in the public setting usually produces *evidence* of effectiveness, not proof. This is because the number of factors affecting an outcome are usually beyond the control of the people in charge of the prevention efforts. In scientific studies, a control group may be established to isolate certain characteristics so that the variables affecting the outcome are known, and the impact of random chance can be eliminated or at least greatly reduced. However, for prevention programs a "controlled" study is not normally possible. Still, gathering data that provides evidence of the impact is elementary and critical to evaluation efforts. It is the foundation on which we build our efforts in the first place, and on which we demonstrate our results.

## TYPES OF DATA

There are several types of data that are important to evaluation efforts. These include anecdotal reports, records, testing, and surveys. This data is a critical part of the formative evaluation necessary for proper program design.

**formative evaluation**
■ the process of gathering information about the scope of the fire problem before design begins, including its root causes and the details about the people who are affected by it

**Formative evaluation** refers to the process of gathering information, before design begins, about the scope of the fire problem, including its root causes and the details about the people who are affected by it. In a simplified version, formative evaluation allows prevention program designers to effectively target their efforts according to fire cause, the high-risk audience, and specific locations.

*Anecdotal* data (descriptions of specific incidents) may be found in newspaper articles, letters from students, interviews, testimonials from organizations, or information offered by individuals. Such data *may* be among the most important pieces of an overall evaluation strategy, because these descriptions of individual incidents help to capture public attention.

The story of a child playing with matches or of people living in an illegal occupancy can sometimes portray the problem more effectively than statistics because it humanizes the situation. A single fire death may not mean a lot when compared with all the others. But a personal story may have much more impact on decision makers because it portrays the human side of the problem. For example, a six- or seven-minute response time to an emergency may not sound like a long time—unless you are the one on the balcony awaiting rescue. Anecdotes providing examples of how individuals are affected by fire can lead to support when we're trying to prevent other losses.

Anecdotal evidence is powerful, but one cannot overlook the need for the larger statistical picture. Individual accounts are important, but one incident (unless it is reflective of a larger problem) may not motivate change or attention to a fire problem. Consequently, records that contain the history of total responses and incidents form the foundation of the statistical analysis that is critical to a more comprehensive evaluation of prevention programs.

Official records that illustrate the nature and extent of fire and life safety problems may be found in a variety of places. Individual fire departments usually maintain a reporting system that provides data helpful in designing prevention efforts. For example, loss statistics for any jurisdiction are generally derived from fire investigation forms, which outline the cause of a fire and the extent of damage. Nationally, many of these emergency incident forms are based on the National Fire Incident Reporting System (*NFIRS 5* is the current version), developed by the National Fire Information Council for the U.S. Fire Administration. Another source of national data is the National Fire Protection Association's (NFPA) "One Stop Data Shop." NFPA provides annual reports and a series of more specific reports that analyze the data they receive each year for others' benefit. NFIRS 5 data is usually compiled at the state or local level, and local jurisdictions can compare their data with statewide information or data from other communities their size. The NFPA reports are actually driven by both NFIRS data and sampling of smaller events throughout the nation. NFPA scientifically extrapolates that data to create a picture (statistical analysis) of the size and nature of the fire problem. Both USFA and NFPA reports provide us with excellent examples of how to display the results of our own data. And because local decision makers usually want to know what is happening in their own area, the local collection and analysis of data is very important. In some areas of the nation, it is mandated by law.

The purpose of these local (or national) incident records is to help define the "who, what, when, where, and how" of fires and other injuries so that they can be prevented. As an example, we obtain that information by collecting data about the number of fires by type (e.g., car fires), by address or location, and by cause (e.g., electrical short). We should also be collecting information about "who" is involved, so that we know where to target our prevention efforts. Knowing the age groups, economic levels, gender, and race of people involved in a given incident, and also whether or not physical disabilities were a factor, would all be important pieces of data to collect. That information is not always available, however, so the critical demographic information about "who" is involved may

# Fires in the United States During 2008

**1,451,500** fires were reported in the U.S. during 2008.

- down **7%** from 2007
- **3,320** civilian fire deaths
- **16,705** civilian fire injuries
- **$15.5 billion** in property damage
- **103** firefighter deaths

Firefighter deaths are not restricted to fires.

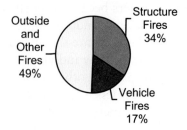

**Fires in the United States During 2008**

- Structure Fires 34%
- Outside and Other Fires 49%
- Vehicle Fires 17%

**515,000** structure fires occurred in the U.S. during 2008.

- decrease **3%** from 2007
- **2,900** civilian fire deaths
- **14,960** civilian fire injuries
- **$12.4 billion** in property damage

**236,000** vehicle fires occurred in the U.S. during 2008.

- down **9%** from 2007
- **365** civilian fire deaths
- **1,065** civilian fire injuries
- **$1.5 billion** in property damage

**700,500** outside and other fires occurred in the U.S. during 2008.

- down **9%** from 2007
- **55** civilian fire deaths
- **680** civilian fire injuries
- **$0.2 billion** in property damage

*The California Wildfires 2008, with an estimated property damage of $1.4 billion is included in the total property damage but it is excluded from the property damage estimates for structure fires, vehicle fires and outside and other fires.

Source: *Fire Loss in the United States During 2008* by Michael J. Karter, Jr., NFPA, Quincy, MA
*Firefighter Fatalities in the United States – 2008*, by Rita F. Fahy, Paul R. LeBlanc and Joseph L. Molis, NFPA, Quincy, MA

NFPA's method for reporting fire loss data provides an example we may wish to emulate at the state or local level. (*Reprinted with permission from National Fire Protection Association, © 2008 NFPA*) require research of other records. Hospital, census, and school records may help more closely define the specific demographics of potential audiences for prevention efforts.

When a medical incident is involved, there may be specific rules about collecting and releasing information that affect our ability to analyze the data. The Health Insurance Portability and Accountability Act of 1996 generally restricts what information can be collected and disseminated regarding burns or other injuries. This privacy act is administered by the Department of Health and Human Services.

Fire inspection records are also a common method of gathering data to help design prevention programs. These records often provide details about the types of hazards found according to occupancy type. That information is valuable when determining what type of proactive efforts should be undertaken to prevent commercial fires (where inspections occur). Inspection records can also be helpful when trying to figure out what training fire inspectors may need. They are also important when trying to determine the workload (or productivity) of individual (or groups of) inspectors.

Public education records are valuable to help determine the number of people contacted and the hours actually spent on public education activity. Often local decision makers want to know how many presentations have been done and how many people have been reached. That information is also valuable to measure overall workload and productivity. However, the impact or outcome of public education programs is more accurately measured in other ways, to be discussed later in this chapter.

## TESTING

Testing is another method of gathering data. Tests can be used to determine the knowledge level of a target audience about (in this case) fire and life safety skills. There are a few critical things that local program managers should know about testing.

First, a test is only a "snapshot" of knowledge levels. To get a true picture of whether prevention programs are raising knowledge levels, pre- and post-tests are necessary to document the actual increase in knowledge. A single test might only reveal knowledge a particular audience already had. Also, constructing test questions requires some expertise. There are instances where a test question asked in a leading fashion can allow the person being tested to guess the right answer without any previous knowledge of the material being tested. Or a question may be written so poorly that no one can get a correct answer. Local decision makers are encouraged to seek professional help in writing tests.

It is the compilation of test scores that provides the data necessary to help evaluate life safety and fire prevention efforts. Scores from tests can reveal individual performance or can be used collectively for a larger audience. More is said about analyzing test data later in the chapter.

## SURVEYS

Surveys, like tests, should be founded in pre- and post-test applications. They are really the same type of data collection instrument as a test, but a survey is generally used for a much wider audience. If, for example, a local jurisdiction wants to evaluate the impact of a prevention program designed for an entire community, a survey is the appropriate evaluation instrument to use. A survey may be oral,

# FIRE & LIFE SAFETY
# CODE INSPECTION

**VANCOUVER FIRE DEPARTMENT**
**FIRE MARSHAL'S OFFICE**
7110 NE 63rd Street ~ Vancouver WA 98661
**360-487-7260** ~ 360-487-7227 fax ~ www.vanfire.org

| BUSINESS NAME | | ADDRESS | |
|---|---|---|---|

### INSPECTION, TESTING AND MAINTENANCE REPORTS NEEDED
☐ SPRINKLER  ☐ FIRE ALARM  ☐ HOOD & DUCT  ☐ HOOD CLEANING  ☐ FIRE PUMP  ☐ GENERATOR  ☐ STANDPIPE  ☐ FIXED CHEMICAL  ☐ OTHER

**NOTICE:** FIRE PROTECTION CONTRACTORS ARE REQUIRED BY VMC TO SUBMIT ALL TEST REPORTS WITHIN **30** DAYS OF SERVICE

**PERMITS REQUIRED**

**NOTICE OF FIRE & LIFE SAFETY HAZARDS:** An inspection of the above-mentioned premises has disclosed the following fire & life safety hazards and/or violations of the provisions of the applicable local or state laws, codes & standards.
**All "CHECKED" section(s) indicate unsatisfactory conditions requiring immediate attention.**

☐ The business representative must call to schedule a reinspection to verify compliance within ____days. A written plan of action must be submitted by fax or email if corrections cannot be completed within the ____day period. Failure to schedule a reinspection or submit a written plan of action can result in an unannounced reinspection at the end of the period and result in additional inspection fees.

| DEPUTY FIRE MARSHAL | | DATE | |
|---|---|---|---|

I acknowledge receiving this report & understand the noted violations are required to be corrected immediately upon receipt.

| OCCUPANT SIGNATURE | | DATE | |
|---|---|---|---|

### HOUSEKEEPING
☐ Combustible material shall not be stored in boiler rooms, mechanical rooms or electrical equipment rooms. (IFC 315.2.3)
☐ Combustible materials shall not be stored in exits or exit enclosures. (IFC 315.2.2)
☐ Outside dumpster shall be kept at least 5' away from combustible walls, windows, doors, overhangs & lid shall be closed. (IFC 304.3.3)
☐ Combustible storage shall be at least 2' below the ceiling or 18" below sprinkler heads. (IFC 315.2.1)
☐ Compressed gas containers, cylinders & tanks shall be secured to prevent falling. (IFC 3003.5.3)

### EXITS
☐ Exit ways & doors shall not be visually or physically obstructed. (IFC 1028.3)
☐ Exit ways & doors shall be unlocked when building is occupied. (IFC 1008.1.8)
☐ Exit signs shall be illuminated. (IFC 1027.3)
☐ Emergency lighting systems shall be functional. (IFC 1006.3)
☐ Fire assemblies shall not be obstructed or otherwise impaired from their proper operation at any time. (IFC 703.2)
☐ Main door shall have a sign above door stating "THIS DOOR TO REMAIN UNLOCKED WHEN BUILDING IS OCCUPIED." (IFC 1008.1.8.3)
☐ Door handles, pulls, latches, locks & other operating devices shall not require tight grasping, pinching or twisting of the wrist. (IFC 1008.1.8.1)

### ELECTRICAL
☐ A working space not less than 30" wide (or width of equipment), 36" deep & 78" high shall be provided in front of electrical service equipment. No storage within this designated work space. (IFC 605.3)
☐ Cube adapters are prohibited. Relocatable power taps shall be polarized or grounded & equipped with overcurrent protection & shall be listed. (IFC 605.4 & 605.4.1)
☐ Relocatable power taps shall be directly connected to a permanently installed receptacle. (IFC 605.4.2)
☐ Relocatable power tap cords & extension cords shall not extend through walls, ceiling, floors, under doors or floor coverings, or be subject to environmental or physical damage. (IFC 605.4.3 & 605.5)
☐ Extension cords & flexible cords shall not be a substitute for permanent wiring. (IFC 605.5)
☐ Electrical wiring, devices, appliances & other equipment that is modified or damaged & constitutes an electrical shock or fire hazard shall not be used. (IFC 605.1)

### COMMERCIAL COOKING PROCESSES
☐ A Class K fire extinguisher shall be mounted within 30' of commercial food equipment using vegetable or animal oils (IFC 904.11.5)
☐ Commercial cooking systems shall be serviced twice a year. (IFC 904.11.6.4)
☐ Hoods, grease removal devices, fans, ducts & other appurtenances shall be cleaned to bare metal. Cleaning shall be recorded, & records shall state the extent, time & date of cleaning. Records shall be maintained on premises. (IFC 904.11.6.3)

Knox boxes are available online at: www.knoxbox.com
By phone: 1-800-552-5669

☐ All fire lanes, hydrants, fire department connections (F.D.C.) or control valves shall be clear & unobstructed. (IFC 508.5.4)
☐ Sprinkler (IFC 903.5) or fire alarm systems (IFC 907.20) shall be serviced annually.
☐ Fire protection systems shall be maintained in an operative condition at all times & repaired where defective. (IFC 901.6)

☐ The building address shall be clearly visible from the street, minimum 4" in height with a contrasting background. (IFC 505.1)
☐ Provide Knox box (IFC 506.1), Provide keys for Knox box (IFC 506.2)
☐ Provide monthly documentation of test on emergency lights (IFC 604.3.1)
☐ Fire-resistance-rated construction shall be maintained. (IFC 703.1)
☐ Provide legible & permanent sign with occupant load posted in conspicuous location. (IFC 1004.3)

☐ A minimum of one 2A-10:BC portable fire extinguisher shall be provided within 75' of travel distance from anywhere in your business on each floor. (IFC 906.3 – NFPA 10)
☐ Fire extinguishers shall not be obstructed & shall be in a conspicuous location. (IFC 906.5)
When visually obstructed, an approved means shall be provided to indicate location. (IFC 906.6)
☐ Fire extinguisher shall be mounted on wall with hanger. (IFC 906.7)
☐ Top of fire extinguisher shall not be more than 5' from floor. (IFC 906.9)
☐ Portable fire extinguisher shall be serviced & tagged annually by a certified individual. (IFC 906.2- NFPA 10)

☐ _____
_____
☐ _____
_____
☐ _____
_____
☐ _____
_____

| ADDITIONAL PAGES ATTACHED PAGE ____ OF ____ | NO PROBLEMS NOTED AT TIME OF INSPECTION |
|---|---|
| ☐ SECOND RE-INSPECTION IN ____ DAYS. | |
| ☐ FINAL RE-INSPECTION IN ____ DAYS. CITATION MAY BE ISSUED IF CORRETIONS ARE NOT COMPLETED. | |
| ☐ ALL VIOLATIONS NOTED CORRECTED ON _____ BY: _____ | |

Inspection checklists are not only a way to keep track of what should be inspected, but they provide fields of data that may be tracked and compared for documentation of safety behaviors—as in hazards that are identified and then abated as part of the code compliance process. (*Courtesy of Vancouver Fire Department, WA*)

| Hazard | Pre-Test | Post-Test |
|---|---|---|
| Flammable Liquid Improperly Stored | 63% | 10% |
| Overloaded Outlets | 18% | 5% |

*Note:* An increased percentage table for certain risk reducing behaviors could be listed separately, yet made a part of the whole table.

| Safety Behavior | Pre-Test | Post-Test |
|---|---|---|
| Smoke Detectors Installed | 50% | 75% |
| Home Escape Drills | 10% | 40% |

Pre- and posttesting is essential to demonstrate that cognitive learning or behavior change has actually taken place in target audiences.

either on the phone or in person. A survey may also be written in the form of a questionnaire. It may be conducted door to door or at places of assembly so that respondents may be randomly selected. In this context, an inspection record is really a survey (done physically by inspectors) to determine how many hazards have been found and abated within a given community. That could be a door-to-door safety survey, with physical observations of the number of hazards found. Or it could be a commercial inspection that yields the same type of information.

Because a survey is conducted to cover a wider audience than a specific classroom, surveys are usually performed with a sample selected from the target audience. *Target audience* in this case means the people for whom the safety message was intended. That could be an entire community, or a specified portion. It is usually impractical to survey an entire community to determine whether they have received and learned from a public education message. Consequently, a sample of the population is chosen to reflect the larger audience.

There are some general steps to a survey process that should be understood. First, the entire population is defined; this is usually an entire community or a select portion of it (such as the inspectable occupancies). It may be all school children or all homeowners in a particular area. Selecting a *random* sample of this larger audience means that each person within that audience has an equal chance of being selected for the smaller sample. This aspect of the random sample implies that the smaller test sample will give results similar to what would be obtained if it were practical to survey the entire audience.

Selecting a random sample, then, requires finding a systematic way of determining who is contained in the entire audience. Ways to do this include voter registration lists, phone books, tax rolls, or utility listings. The next step in the process is selecting an interval. For example, if the larger audience is 50,000 people, a proper interval might be every 500th name on the list, yielding a sample size of 100 survey respondents. Generally, it is best to use the largest sample that one can afford because it reduces the chance for error.

After the interval is selected and names are identified, the survey instrument must be designed, distributed, collected, and analyzed. Once again, local authorities should consult professionals about the actual construction and administration of a survey instrument, because the potential for getting erroneous information is great.

However the data is collected, it must be analyzed before it can be meaningful when we are planning our programs in the first place, or evaluating them afterward.

## Survey Process

### Identify Sample Population

What population . . .

School age, elderly, residents (community, census tract), homeowners City, County, State, Region

### Select Systematic Sampling

Need a list of the population - phone book, utility billing list, voter registration, tax rolls.

### Determine Interval Selection

How many interviews can you afford? Consider time, money, staffing.

Large samples mean less statistical error, use the smallest interval you can afford.

Example: Population = 50,000
         Interviews =    500
          Interval =     100   select every 100th name

### Construct Questionnaire

Brainstorm, keep it simple, ask few questions, provide answers Pre-test the questionnaire

### Collect Data

Conduct pre-test, conduct educational campaign, re-sample population, conduct post-test.

### Analyze, Summarize Findings

Tabulate results, chart or graph findings, statement of analysis, outcome.

# Data Analysis

Most have heard the old saying that "figures don't lie, but liars often figure." It is a useful cautionary note about analyzing information collected for purposes of evaluation. Two people using the very same statistics may draw different conclusions. Consequently, great care should be used when analyzing information. It should be no surprise that many jurisdictions hire analysts with professional training, education, and sometimes certifications in data collection and analysis. Data and the conclusions drawn from it may (in fact should) be challenged, so professional assistance is highly recommended. But for those who cannot afford professional help, or who need to know enough about it to select someone qualified, a few simple rules can help guide our analytical efforts.

## PRE AND POST INFORMATION

As previously stated, information that illustrates preconditions and postconditions can be used to analyze program effectiveness. This information can be evaluated

**TABLE A**

QUESTION 4

| | PRE-TEST | POST-TEST |
|---|---|---|
| CORRECT | 11 | 21 |
| INCORRECT | 18 | 4 |
| | 29 | 25 |

In pre- and post-test or surveys, items may be analyzed by individual numbers or by percentages. Caution is needed so as not to misinterpret or misrepresent the data by manipulating either.

Table A shows that the majority of students did not know the correct answer to Question #4 when the pre-test was given; however, when the post test was administered, the majority of the students responded correctly. Notice that the pre- and post-test samples are of unequal size. In refining our analysis, a table of percentages was constructed. This table dramatized the actual difference in the proportion of correct responses comparing pre- and post-test samples while taking into account the error factor of differing class sizes.

**TABLE B**

QUESTION 4

| | PRE-TEST | POST-TEST |
|---|---|---|
| CORRECT | 38% | 84% |
| INCORRECT | 62% | 16% |
| | 100% | 100% |

for specific questions or topics, or more general views can be revealed. Generally, it is better to convert the data and interpret them in terms of percentages, which can better reflect what has then taken place for target audiences as a whole. For example, it can be useful to know how many people had smoke alarms installed both before and after a specific campaign. That would be a single-item analysis. But if the collective test scores before and after the program are known, the percentage of change may ultimately be more revealing.

As a cautionary note, when the data is summarized, it should be carefully screened for bad information (such as incomprehensible or blank responses). The data should all be placed in the same format, and it is a good idea to make sure that each response can be coded in case computer analysis becomes possible. "Bad" information can invalidate a survey or test; therefore, professionals usually make some effort to obtain missing results, though that is rarely practical. Of course, the level of attention to detail will be determined by the type of sophistication one can afford—but it is good for practitioners to understand that the statistical validity of the pre- and post-testing instruments can be challenged, and therefore the conclusions drawn from the evaluation instrument may be challenged as well.

## SINGLE AND SUMMARY ITEM ANALYSIS

Tests and surveys can be analyzed by each item or by a summary of items. That is one of the reasons why it is valuable to express answers in terms of percentages. See the accompanying example.

Despite what was noted earlier about summarizing test or survey results, persons analyzing information should also understand that percentages can often be

## SINGLE AND SUMMARY ITEM ANALYSIS

As discussed in GATHERING INFORMATION, tests and surveys can be analyzed by each item or by the summary of items. In any case, it is true that the raw data should be interpreted in terms of percentages to better indicate the changes and eliminate problems with different sample sizes.

Results arrange in a tabular format produce an efficient visual summary and lend themselves to the application of statistical tests.

To produce an evaluation of the effect a particular campaign has had upon the knowledge level skills and/or attitude(s) of a population there should be measures of the knowledge level skills and/or the attitudes of the population at two points in time: (1) before the education campaign has begun (the pre-test), and (2) after the campaign has been conducted (the post-test). Having sampled the population at these two times, the results can be placed in a series of tables for each question.

Example: Pre- and post-test on smoke detector maintenance

| Question #1 | Pre-test | Post-test | % Difference |
|---|---|---|---|
| CORRECT | 38% | 84% | 46% |
| INCORRECT | 62% | 16% | |

| Question #2 | Pre-test | Post-test | % Difference |
|---|---|---|---|
| CORRECT | 25% | 80% | 55% |
| INCORRECT | 75% | 20% | |

*Note:* Questions can be listed by number, and referred to in additional text, or they can be listed by question content.

Example:
Question: Smoke Detectors should be tested **monthly**.

| | Pre-test | Post-test | % Difference |
|---|---|---|---|
| CORRECT | 25% | 80% | 55% |
| INCORRECT | 75% | 20% | |

misleading. Expressing a position that fire deaths have fallen by 50 percent in 1 year may sound dramatic, until those reviewing the actual data determine that it was a fall from two fire deaths to one. Even under the best of circumstances, this would not be considered statistically significant. In this context, statistical significance means that the changes occurred more than likely by some effort, and not because of random chance. That element of random behavior can cause fluctuations in numbers without any other factor. One year fire rates may be up, the next down, through no intervention efforts of our own. Statistical significance is a topic that requires specialized training so that we may be able to defend the conclusions we draw from our analyses.

## TREND ANALYSIS

If measuring percentages can be misleading, so too can looking at changes that occur only once. Who is to say that a reduction in fire deaths did not occur by random chance, rather than because of our prevention efforts? Trend analysis can be used to measure the impact of program efforts over time and statistically validate differences from previous trends. A simple version of a trend analysis is

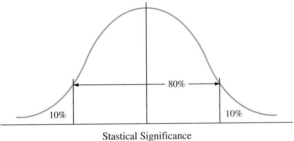

Stastical Significance

- Normal variances occur
- Apx. 80% within two standard deviations from the mean average
- Changes outside standard deviations are likely significant – not due to random chance

Changes that occur farther outside the statistical norm for variation lend credence to claims that the changes did not occur as a result of random chance. They provide evidence, not proof, of impact or outcome.

offered here to demonstrate the process and its use. Charting information over a period of time gives some indication of the normal fluctuation from year to year. The point of a trend analysis is to see if program efforts are changing the course of the data in a significant way. In simple terms, it works as shown in the accompanying figure.

Generally speaking, the farther the data points fall away from the mean average difference area of a projected trend, the more statistically valid is the claim that the change is the result of the specified prevention campaign or program. Smaller numbers have a greater chance for error, and there are almost always outside factors that could be influencing the changes we observe even when we note them over a period of time. We do not usually have the resources to conduct formal studies where control groups can be established to rule out outside influences, allowing scientifically solid conclusions to be drawn. Because of the number of variables involved, caution is needed when evaluating data trends and making statements about the conclusions drawn. And even though modern computer programs can produce a statistically valid trend analysis, seeking professional assistance is important to ensure the validity of the evaluation.

Trends may also be used in a comparative fashion. Trend analysis graphs can be used to compare current efforts with previous trends, with other jurisdictions, and with state or national data. The point is to determine, by examining the trends, whether program efforts are having any impact on the problem.

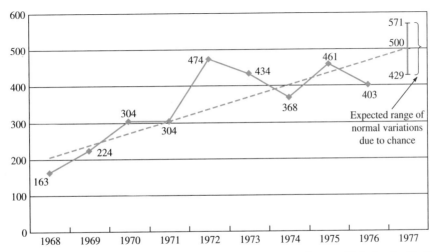

Trend analysis is used to evaluate results over time to avoid making mistakes about changes that may be the result of random chance rather than educational efforts.

Trend analysis may be used to compare a jurisdiction's numbers over time—to itself—or to compare one jurisdiction to another. This may reveal whether changes are a result of random chance or normal trends, or whether educational programs are producing results. Once again, this analysis provides insight and evidence—not proof—of a connection between the numbers and prevention efforts.

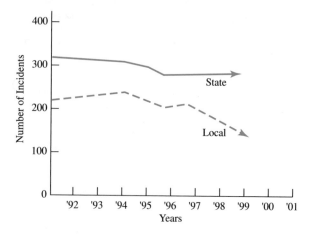

Finally, the person responsible for gathering and analyzing data will ultimately have to present it to decision makers. The presentation of data is very important. A variety of charts and diagrams can usually tell the story far better than any verbal presentation. A combination of personal anecdotes *and* statistical data can be used in a highly effective presentation. The illustrations shown in this chapter provide some examples of how data can be arranged and presented in meaningful terms.

# General Versus Specific Performance Measures

The different types of general performance measures have already been described to a certain extent. They have been listed as general because they are used by organizations such as the Government Accounting Standards Board, the International City Managers Association, and the Association of Government Accountants to examine the performance of different aspects of government. They tend to examine comprehensive efforts, rather than specific programs. For example, government auditors have an interest in how much funding prevention programs have received (collectively), how efficient they have been, and what kind of results have been achieved. Auditors tend to measure workload, efficiency, and effectiveness of data for whole prevention programs, such as all code enforcement or all investigation efforts.

Program-specific evaluation measures also look closely at results, but tend to focus less on workload or even efficiency, and instead look closely at the outcomes (results) specific programs achieve. For example, a public education program designed to promote working smoke alarms would be of interest to both government auditors *and* prevention practitioners or researchers. But the auditors want to know the costs for *all* prevention efforts, not just the smoke alarm program. They want to know about fire investigations, code enforcement, plan review, and other public education efforts. The prevention professionals or public safety researchers with an interest in demonstrating what works, and what does not, will delve more deeply into a single program with more comprehensive evaluation measures.

In this context, then, the difference between the two terms is that performance measurement can document that overall programs are providing evidence of results, whereas evaluation measures refer more often to specific prevention programs.

## MODEL EVALUATION MEASURES

An ongoing effort to develop model fire prevention and life safety program performance measures may be found at www.strategicfire.org. That effort is a collaborative approach to providing a consensus on what the performance measures should include. By establishing model evaluation measures for prevention programs, practitioners and researchers in the field have tried to develop terminology that can be replicated throughout the nation. The goal of those in the prevention field is to begin reporting fire prevention efforts in a consistent enough fashion to allow for legitimate program comparison and the establishment of both baseline performance measures and benchmark standards. In other words, program managers can compare results to their own history, or to other jurisdictions, to begin formulating management decisions based on evidence. The establishment of consistent and accepted performance measures will allow for demonstrated evidence that prevention programs are producing a desired result. They can also provide evidence that programs may not be working, and therefore require modification.

However, there are some strong cautionary notes. The number of potential variables and the complexity involved in establishing model performance measures for fire prevention programs has often proved problematic.

Generally, the results we expect from fire prevention (and related life safety) programs include documenting educational gain (people actually learning something as opposed to just sitting in class); documenting risk reduction where increased safety behaviors or decreased hazard-producing behaviors can be documented (e.g., hazards noted and abated during fire code compliance inspections); and finally, documented reductions in losses. *Losses* in this context means reductions in deaths, injuries, and both direct and indirect economic losses.

But as mentioned before, there are often extraneous factors that can affect the data received and its perception. So although prevention programs exist to increase safety knowledge and to reduce risks and losses, decision makers should be cautious about abandoning efforts (as some have) as a result of inadequate data or a poorly designed study. For example, numbers for incident rates may rise, fall, or remain constant because of prevention efforts, random chance, data entry errors, or (in the case of wildland fires) because of the weather.

For that reason, most of the model measures here have been described as changes, and the act of beginning to evaluate performance of prevention programs in common terms is the goal. With field experience over time, evaluation methods will increase in sophistication, and program managers can begin to compare the results of their efforts in more scientific terms.

For comparative purposes, we have grouped these result-oriented performance measures into common evaluation terms used and taught at the National Fire Academy, the National Fire Protection Association, and other organizations. They include (for postprogram evaluation) process, impact, and outcome evaluation measures.

Outcome, Impact, and Process measures are the recommended terms for comparative analysis of the results our prevention programs achieve. This chapter contains examples of each for the basic functions of a fire prevention effort. That includes plan review for new construction, code compliance inspections for existing businesses, fire investigation, and public safety education activities.

## TERMINOLOGY

The difference between the following terms is actually driven by the stated goals of the particular program being evaluated. But generally, they are defined in the following fashion relative to fire prevention programs.

**Outcome evaluation:** the mechanism of determining how well a program achieves its ultimate goals (such as reduced losses).

**Impact evaluation:** the mechanism of measuring changes in the target population that the program is intended to produce. These measures could be considered advance indicators of successful outcomes (such as reduced risk).

**Process evaluation:** the mechanism of testing whether a program is reaching its goals, such as reaching target populations with quantifiable numbers expected. Process evaluation measures often include measures for workload (e.g., number of inspections done per inspector) and milestones in achievement of process objectives.

Measuring workload does not indicate success of a program's results. However, it can be important to decision makers when deciding how many resources are required to perform certain prevention tasks, such as inspections, plan reviews, investigations, or public education activities.

The following examples are intended to *stimulate* the concept of common performance measures, so that comparisons between like organizations are possible in an "apples to apples" sense. *They are not intended to be a complete listing,* but rather provide a beginning methodology for comparative analysis of the *results* prevention programs achieve. Many of the measures have cautionary notes directly attached to help clarify the intent of the measure.

The model evaluation measures do include some of the same types of measurement found in overall performance measurement. Process measures, for example, may provide insight into the workload involved in a particular prevention program. But the real overlap between the two approaches to evaluation occurs

---

**outcome evaluation**
- the mechanism of determining how well a program achieves its ultimate goals

**impact evaluation**
- the mechanism of measuring changes that the program is intended to produce in the target population

**process evaluation**
- the mechanism of testing whether a program is reaching its goals

---

These examples of model evaluation measures apply to the various prevention programs found in a comprehensive approach. They provide some insight into what formative, process, impact, and outcome measures might look like for each type of program. More information may be found on this topic at www.strategicfire.org.

**Stages of Evaluation**

- Formative
- Process
- Impact
- Outcome

when we are trying to describe the results of our efforts. At this point, whether it is impact or outcome, we may describe the results we achieve in some fairly specific ways. Some have already been mentioned and cautionary notes applied to each type. Here each is described in a bit more detail.

## RESOURCE CHANGES

One of the most often overlooked measures of a program's impact is the number of resources it has generated. Most cost/benefit analyses take into account program costs and the resources necessary to accomplish a program's implementation. However, some types of prevention programs produce results that are not anticipated.

For example, a public education campaign might generate contacts with the television industry that result in an in-kind contribution of free television time. This might not be caught or reported unless we are consciously tracking the resources a given program may require *and* produce. In this example, media time must usually be purchased if fire departments want to get their information out during peak hours, but television and radio stations will often offer "two for one" purchase packages that net a greater result. Tracking, documenting, and taking credit for the resources obtained to support prevention efforts may be one of the more impressive aspects of its evaluation for those who control the purse strings.

These resource changes are then part of the results we may want to document and report—whether it be for overall performance measurement or program-specific evaluations. These changes in resources are usually described as a process evaluation or an impact a program may achieve, as opposed to an outcome.

## EDUCATIONAL GAIN

Educational gain is a change in the knowledge level or attitudes held by the people identified as the target audience of a particular project or campaign. Individually or collectively, testing or surveying people will be required to document the changes in their knowledge levels. An understanding of how people learn is the foundation of this aspect of prevention efforts.

Simply put, people learn in the cognitive, affective, and psychomotor learning domains. *Cognitive* means that they have grasped the information in a tangible (documentable) way. They can recite the proper steps about what to do in the event their clothes catch fire. For example, the foundation of NFPA's *Learn Not To Burn*® program is teaching a specific action for a specific problem. Recitation of a correct response is an indicator that people have grasped the message (learned it cognitively). However, it does not mean they will necessarily apply it.

The effective learning domain gets at the attitude or motivation of the learner. Using the previous example, a person may understand "stop, drop, and roll," but they may feel that actually learning it is childish and, therefore, never learn it completely. If their attitude is such that they will not actually perform the task, then the level of learning never reaches the point where behavior is actually changed. So the physical aspect and performance of the new learning is very important and is driven in part by the attitude of the learners.

So the final learning domain is an indicator that they will actually be able to perform the task if it is required.

The psychomotor learning domain involves physical performance. The same group that thinks "stop, drop, and roll" is for kids might not be motivated to demonstrate that learning by actually performing it in a test of learning. They are not properly motivated and, consequently, the instructor may never know if they were actually learning the physical requirements of the task. Some difficulties with the physical application may not be evident until the maneuver is actually tried.

In simple educational terms, the meaning of these learning domains is: People learn from what they hear; they learn more from what they see; they learn still more from what they do; but they learn best from what they hear, see, *and* do. Applying these learning domains to document educational gain leads to the next type of measure, risk reduction.

## RISK CHANGES OR REDUCTIONS

Risk reduction is one of the more tangible aspects that measure a prevention effort's effectiveness. It means that the evaluator can document that the education was applied in the real world. Numerous examples include a variety of hazard-reducing or safety-increasing actions that can be documented, summarized, and presented as evidence of program impact.

For example, the entire process of conducting fire code enforcement inspections is an example of risk reduction. Each inspection is an opportunity to document the number of hazards found and abated through the inspection activity. Eliminating hazards such as improper extension cords or burned-out exit lighting systems is ultimately an example of how the overall risk for that occupancy has been reduced. Even if only for a short time, decision makers have a tangible example of the impact prevention efforts produce.

Other ways to document risk reduction for any particular community do exist. Home safety inspections are an example of voluntary activity on the part of individuals to reduce their hazards or increase their personal safety. For example, people might remove hazards, such as gasoline stored for lawnmower use, from the basement of their home and place it in a safety can within their garage or an outbuilding. This would be an example of hazard-reducing activity that reduces risk. Additionally, they might increase their safety activities to help minimize their overall level of risk.

Examples that lend themselves to documentation include:

- The number of smoke alarms installed before and after a program designed to encourage their installation
- The number of families conducting home escape planning drills after a community-wide campaign to promote them
- The number of families that empty ashes from their fireplaces in an approved fashion after a campaign designed to teach them the proper methods for doing so

In each case, an assumption is made that the information obtained can be documented and is reliable. If observations are made in person, such as in a door-to-door survey or code enforcement inspection, the information is generally more reliable. However, tests or surveys can be performed to illustrate the relative risk reduction for a community. It is necessary to be cautious because of the phenomenon in which people give answers that they believe interviewers want to hear, rather than speaking truthfully.

As previously stated, documenting changes in risk is more realistic than focusing solely on the term *reduction*. This is important because some decision makers have abandoned prevention efforts without adequate study, based on preliminary findings that physical safety behaviors were not changing.

This aspect of specific measurement is most often described in terms of program impact. That is because it does not always lend itself to a reduction in losses—but does provide evidence that tangible results (in this case impacts) are being achieved.

## LOSS REDUCTION

Loss reduction is the most basic and the most important effectiveness measure for prevention efforts. It is the outcome we desire. Ultimately, the motivation of any prevention program is to produce results that reduce losses or save lives. Expressing the results that prevention efforts have in reducing dollar losses from fire or other incidents, property losses, and the actual numbers of those incidents is another way of measuring loss reductions. Making sure the information is valid and presenting it in a fashion that is not misleading is imperative. A trend analysis is the best way to make sure that the data is not being misconstrued. Examples of loss reduction include:

- Hospitals and clinics in a community report a 40 percent reduction in burn incidents over a 5-year period following burn prevention programs sponsored by a local coalition of fire departments, insurance companies, hospitals, and public health agencies.
- After increasing education for students in their community, incidents of false alarms are reduced by 37 percent over a 4-year period, where the previous trend showed an overall gain in incidents of 95 percent.
- An overall reduction of fire incidents of 60 percent was documented over a 5-year period after education programs were designed and implemented in schools and at community events. Previous trends showed a much smaller reduction.

As with risk reduction, strong cautionary notes need to be applied to false conclusions where the data on losses are concerned. Normal (random) fluctuations in the numbers will always occur. There may be external factors that affect the numbers of incidents we observe, regardless of our prevention efforts. Here again, looking at the *changes* in losses might reveal that we need to make adjustments in our programs, rather than eliminate them outright—even though our stated goal is to produce an outcome of less loss from fire.

## BENCHMARKING

According to William Gay of the Public Management Group, "Benchmarking is the surrogate for the competitive forces that continually push businesses to achieve higher levels of quality and productivity—productivity that serves the business, its employees, and its customers."[1] Originally developed in the private sector, this method of evaluating the impacts and outcomes of prevention programs may be among the most powerful tools we have, because it may provide the closest thing we will see to "control" studies where differences between programs can be isolated. Gay maintains that benchmarking is particularly appropriate for public agencies because of the willingness with which public agencies share information. In a way, benchmarking is a search for the best practices that lead to superior performance, and creating our common comparative data fields provides us with a way to compare in an "apples to apples" fashion.

As such, benchmarking could be done for each type of performance measure. Workload, efficiency, and effectiveness measures could all be compared with other jurisdictions to see which programs are producing the best results. Whether or not formative evaluation efforts were undertaken could be compared as well. Certain process evaluation methods could be compared—like to like. And the overall impacts or outcomes (such as risk reduction or loss reduction) could be compared in a benchmarking process.

In similar fashion, benchmarking can be done by individual communities that wish to compare themselves to themselves. For example, a local community might set a goal of reducing fire incidents by 30 percent in a particular period of time. Using trend analysis, they could demonstrate whether the goal had been achieved and whether it was statistically significant given previous trends. As Gay quotes in his report, Robert Camp of Xerox Corporation summed up benchmarking and its importance in the following way: "The process of searching out and emulating the best can fuel the motivation of everyone involved, often producing breakthrough results."

Once again, care should be given to make sure that jurisdictions really are comparing "apples to apples." Many variables can affect the results, including weather, demographics of the community, the age of buildings, and the previous history of a particular community. That is why some find benchmarking a valuable tool to compare one part of a city with another or one fire station's prevention efforts to another within the same jurisdiction. Ultimately, benchmarking uses the same data gathering and analysis techniques as the other methods of evaluation, but does so in a comparative fashion.

# Summary

Evaluating prevention efforts can be done a variety of ways. General performance measurement (programs overall) or program specific evaluations share some common measures, but the goals of each are slightly different. Government auditors usually want to know how overall programs perform, so they look at certain types of measurement tools. Workload measures reflect how much work is being done. Efficiency measures refine workload measures to indicate whether or not they are being done in the most efficient way. Effectiveness measures demonstrate the impact or outcome of a program. The results our efforts achieve are also of interest in program specific evaluations. So if the common measurement methodologies between general and program specific are about results, the specific measurement terms we use to describe those results would work for both approaches.

The effectiveness measures most commonly used include educational gain, risk reduction, and loss reduction measures. Measuring educational gain provides evidence that public fire and life safety education activities are producing the desired learning result. Pre- and post-testing practices can document whether the recipients of an educational program are actually learning or merely sitting through a presentation they will forget the next day.

Reducing risk is one of the basic reasons for having a comprehensive prevention program. Measuring risk reduction means documenting an increase in safety behaviors or a decrease in hazard-producing behaviors. For example, documenting the number of working smoke detectors in a community can provide evidence that the risk of dying in a fire is reduced. This is true because national statistics indicate the effectiveness of smoke detectors in saving lives. A compilation of hazards abated during fire inspections can also provide evidence that risks have been reduced because hazards have been removed.

Measuring risk reduction is a valuable part of an evaluation strategy because it provides some quantifiable indicators that can be used to determine the impact of a prevention program. Local decision makers can quantify how many people have working smoke alarms and how many practice fire escape planning. They can also quantify how many community fire hazards have been abated in a code enforcement inspection program.

However, the ultimate performance measures will always be those that document loss reduction. Workload measures demonstrate that employees are doing an adequate amount of work. Efficiency measures provide some indication of how quickly things are done or what they cost in relation to other similar services. However, measuring the reduced losses that occur as a result of prevention programs will provide the most evidence of positive results. This is the ultimate performance (or outcome) measure that justifies the expense of conducting prevention programs. However, local decision makers should be cautioned against leaping to conclusions based on short-term analyses of loss data. *All* the performance measures looking at effectiveness should be evaluated over a period of time, because the change in activity can be caused by normal variations in any statistical analysis. Looking at loss reduction over a period of time can provide the best picture of whether a local jurisdiction is improving or not. In addition, it is more accurate (statistically) to compare any jurisdiction to its own history, rather than trying to compare it to another. However, benchmarking has become a valuable way of identifying best practices and comparing efforts against them in an effort to improve performance. Ultimately, the purpose of evaluating a program is to improve it.

# Case Study

## Porth Bureau of Fire, Rescue, and Emergency Services

The City of Porth has a population of about 500,000 and serves an area of about 150 square miles. The fire department has 27 fire stations and about 750 sworn and nonsworn employees, and it provides a full range of emergency and prevention services. The Prevention Division has about 70 employees devoted to public education, fire investigation, code enforcement, and plan review activities. Over time, the efforts of individual programs have been evaluated in a variety of ways.

Workload indicators are tracked for the city auditor, who produces a regular report on the activities of the fire department. The annual report on service efforts and accomplishments is shared through the International City Managers Association and the Government Accounting Standards Board.

Efficiency measures have been tracked in general terms and used in a benchmarking process with other jurisdictions. Effectiveness measures have been tracked closely and documented for specific programs such as public education campaigns. Overall, the measures demonstrating risk and loss reduction have dominated the discussions about the relative value of prevention programs in Porth.

Recently, a new program of charging fees for fire code enforcement inspections was instituted. As part of that program, a 5-year evaluation period was established to demonstrate the impact of the program. The measures for the program include risk reduction and loss reduction measures, but they also include more specifically targeted measures for the occupancies subject to the code enforcement activities. Several examples of the data analysis are shown in the accompanying figure.

# Case Study Review Questions

1. Given the information presented, what measures portray risk reduction and loss reduction in the Porth case study?
2. What inferences can be made from the comparison between Porth and other comparable jurisdictions in the Service Effort and Accomplishment (SEA) report? What dangers are there to these comparisons?
3. What conclusions can be drawn from the comparison of Porth's fire loss statistics to state and national trends over the same period of time?
4. What indicator in the Porth case study reflects workload? What factors might affect the workload indicators to produce these results?
5. Where are examples of anecdotal instances in the case study? How much of an impact (presumably) would these indicators have on the decision makers responsible for continued funding of the fee for inspection program?
6. How is evaluation linked to planning?
7. How may we describe results of fire prevention programs?
8. What is the value of benchmarking?
9. What is the difference between performance measurement and evaluation measures?
10. How does the term *results* overlap both performance and evaluation measures?

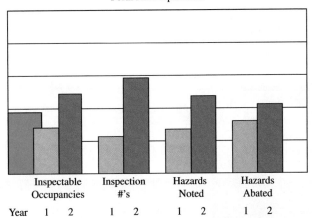

Porth Fire Department

These graphs illustrate the results achieved by the new inspection program begun in Porth.

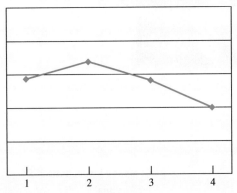

Trend of Fires by Inspectable Occupancy

# Endnote

1. Gay, William G., *Benchmarking: A Method for Achieving Superior Performance in Fire and Emergency Medical Services.* Herndon, VA: Public Management Group, 1992.

# 13

# Organizing Fire Prevention Programs: Staffing, Training, and Funding Options for Prevention Programs

## OBJECTIVES

After reading this chapter you should be able to:

- Identify several options for staffing and funding life safety and fire prevention programs.
- Describe how fire prevention fits into community efforts.
- Describe the importance of leadership in fire prevention efforts.

Finding the right person for any job can be a difficult task. Given the broad nature of prevention programs—including fire investigations, plan review for new construction, fire code enforcement, and public education—finding the right person for all these tasks may be nearly impossible. Decision makers are often forced to rely on whoever is available, rather than whoever would be desirable for the job.

However, it is essential to have qualified personnel working on prevention programs, if for no other reason than to protect the jurisdiction from legal liability for jobs done poorly. A "one size fits all" approach may not be effective. In fact, the various tasks that must be performed may each require a different skill set. Good presentation skills and knowledge of educational methodologies are obvious requirements for someone involved with public education—but we would be looking for something different in a fire investigator. Good analytical skills would be among the traits we were seeking in that case, as well as a knowledge base in fire investigation techniques. More detail about specific job functions will be provided later, based on a history of job performance analysis conducted for professional qualification documents at the National Fire Protection Association. But first, a more general discussion is appropriate.

# Staffing Options

Fortunately, there are a variety of staffing options to cover prevention activities of all types. Whether for plan review, code enforcement, investigations, or public education, the mix of staffing options stems from two basic sources: paid personnel or volunteers. However, after that decision is made, the mix of how they are deployed becomes almost infinitely more varied, and how they are managed throughout the nation also varies considerably.

Many jurisdictions, such as Seattle, Washington, use emergency response personnel to conduct all or part of their fire code inspections. Commonly, this type of code enforcement activity is called a **company inspection**, because it uses fire and medical emergency crews (called companies) to conduct the inspection.

company inspections
■ when on-duty emergency response personnel conduct fire code compliance inspections

## AUTHOR'S NOTE

Company inspections are used widely in many fire departments, but there are significant challenges to successfully implementing them. A great deal depends on how the firefighters are trained. Anecdotal evidence among fire service leaders over the years has demonstrated the difficulty many have in getting firefighters to embrace prevention.

The people who aspire to be firefighters are not usually the same ones who are drawn to prevention efforts. The thrill of fighting fire is an attractor—and prevention efforts just do not provide the same level of excitement. And if the emergency responders are not properly trained to conduct prevention activities, the negative aspects can multiply. First, if they feel uncomfortable with the job because of a lack of knowledge, their performance will suffer. Many in the prevention field refer to "drive-by" inspections, meaning a job that was poorly done because of a lack of training.

Combine that problem with a lack of motivation toward prevention in general, or a lack of oversight, and firefighter-based prevention programs can languish and even die.

Current research being conducted by the National Fallen Firefighters Foundation is attempting to identify proper recruitment and training methods that will enhance the ability of fire departments to instill a prevention mentality into their workforce.

There are other departments that use a combination system. For a time, the Portland Fire Bureau in Oregon used a combination of regular inspectors (on a 40-hour workweek) with 53-hour emergency response shift inspectors, who also served as the fifth person on a truck company. This fifth position on an emergency response company conducted code enforcement inspections during the day and responded as the fifth person on the firefighting apparatus at night. This program was discontinued when a local ordinance authorizing fees for inspections was passed, and regular 40-hour inspectors were restored to the budget.

Some fire departments are expanding their code enforcement options with self-inspection programs, in which businesses are asked to identify their own hazards and abate them voluntarily. Self-inspections are often the only substitute for prevention programs that lack the resources to inspect their commercial occupancies on any type of regular basis. In Arizona, the cities of Mesa and Tucson use self-inspections extensively for lower priority business occupancies, defined as the types of businesses that generally have fewer fires but are also relatively simple inspections. Both organizations use the program to gain valuable business contact information for emergency responders, and both find some

Many jurisdictions use emergency responders for fire code compliance inspections—but there is some debate in the field about the quality of those inspections, the workload they place on emergency responders, and the overall results they achieve compared to fully designated inspection staffing. (*Courtesy of McKinney Fire Department, TX*)

value in obtaining current (annual) business information for other operations (such as business licensing) in their respective cities. Tucson initiates a regular code compliance inspection for every business that does not turn in its self-inspection form, as an incentive for local businesses to comply with the self-inspection program. Mesa is currently still evaluating their program to see if such an incentive will be necessary.

Some volunteer departments also use volunteer inspectors, as in the case of Nassau County in New York. They are unusual in this regard, because the most widely used staffing options involve either paid specialty personnel or use of on-duty emergency responders. These are the most frequently seen options for fire code compliance inspections, but there are other fire prevention activities with their own staffing options as well.

## AUTHOR'S NOTE

There is no known study that stipulates which type of inspection is most effective, but evidence provided by an older study[1] published by the National Fire Protection Association (NFPA) suggests that the frequency of inspections does have a correlation with a reduced fire loss rate. More recent research on the topic lays the foundation for further study about the effectiveness of code compliance inspections and which staffing options work best. The results of that research are not currently available on NFPA's Web site, but anecdotal discussions with those involved in the study still do not answer questions about which approach to conducting fire code compliance inspections (company inspections vs. dedicated trained staff) works best. We tend to think of code compliance inspections as confined to commercial occupancies. However, some local jurisdictions in areas prone to wildfires have adopted codes related to creating and maintaining defensible spaces (and other safety features) for homes in urban/wildland interface areas. Code compliance inspections for these properties are much more common in southern California. In this regard, Los Angeles and Los Angeles County fire departments have taken the concept of self-service to another level. They have in the past developed programs to train citizens how to protect their homes from wildland/urban interface fires. Training people to created defensible spaces around their homes is fairly common—but training them to actually stay in place and protect their homes is also done in some jurisdictions as well. This approach to citizen fire protection is not without some controversy. Australia has developed a "stay and defend" approach to wildland firefighting. A recent series of serious and fast-moving fires killed a number of residents, calling the practice into question. More on the topic of defending in place and current refinements to that methodology can be found by visiting the Web site of the Fire and Emergency Services Authority of Western Australia.

More common uses of volunteers for inspection and other activities occur in Europe and Japan, where cadres of citizen volunteers help in times of emergency, but also serve as a critical part of the public education efforts for their prevention programs.

Similarly, most of the United States is still protected by volunteer fire departments, using personnel who do both emergency response and fire prevention activities. These people are not necessarily recruited specifically for volunteer work in prevention; more commonly, they are used as combination personnel with an emphasis on emergency response.

Many departments use firefighters as inspectors, investigators, or public educators, or promote them into a specialized unit to conduct the department's fire prevention programs. Other jurisdictions use personnel hired from outside the fire department and recruit employees with specialized education and experience relevant to their assigned tasks. Mesa, Arizona, for example, uses the personnel within their building department to conduct plan reviews for life safety and fire issues. Others hire fire protection engineers to work within the fire department to assist with or conduct their review of plans for new construction.

Some fire departments coordinate efforts of a variety of agencies in order to achieve their prevention goals. The Houston, Texas, fire department developed a *Tri-Ad* inspection program that involved a partnership with their building department and the local Building Owners and Managers Association. Doing so allowed them to maximize their resources and limit complaints from the business community about multiple code enforcement inspections. There are some departments that do not conduct fire inspection activities at all and refer them to another agency—usually a building department—for fire code compliance.

Many departments also rely on their emergency responders, paid staff, or volunteers for public education and fire investigation activities. Some jurisdictions offer voluntary inspections for private homes and have demonstrated measurable reductions in home fire losses as a result. This type of program is receiving a great deal of emphasis in the United States as one of the types of prevention activities with impressive, documented results. It is difficult to find source documents now—but Edmonds, Washington, in the early 1980s (according to TriData), conducted home safety inspections regularly using paid civilian personnel. They achieved a 62% reduction in fire incident rates for the two years they conducted the program—and when it was eliminated because of funding cuts, the fire incident rates rose to previous levels.

Currently, in the United Kingdom and other nations, major efforts are underway to provide home safety inspections and smoke alarm checks for high-risk homes in their jurisdictions. The results achieved by these programs, and reported by TriData and the Centers for Disease Control and Prevention, are providing impetus for testing similar concepts in the United States.

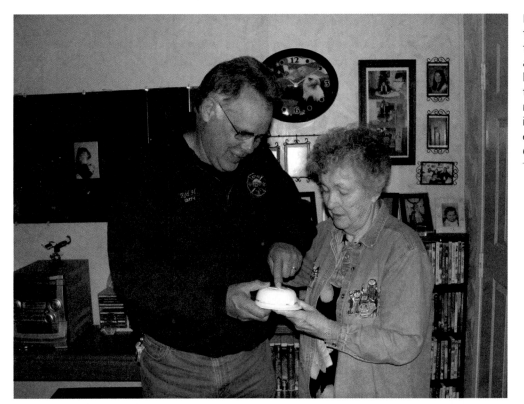

Other departments have their emergency responders conduct initial fire investigations, calling in dedicated staff for larger events. Some departments rely on neighboring police jurisdictions to conduct the investigation when arson is suspected. In other cases, the paid investigators who work for the fire department are also sworn police officers so that the continuity of the investigation is coordinated "in-house."

## POTENTIAL PROBLEMS IN STAFFING

Some departments have a mixture of civilian and sworn fire prevention personnel. Others, such as Vancouver, Washington, hire their fire prevention personnel exclusively as civilians from outside the fire department. Many jurisdictions have had difficulty integrating these approaches, with cultural barriers often springing up between sworn and nonsworn personnel. Many firefighters tend to denigrate fire prevention and those who perform those duties as being less important to the fire department mission. Some prevention personnel who specialize in that field in turn denigrate firefighters as undereducated "brawn" for a blue-collar job. As a result, an organizational cultural separation can arise between the groups, even resulting in open hostility. And during budget cuts, the two groups can become adversaries.

There is no single solution to this problem anecdotally shared by many practitioners throughout the nation. However, efforts to bring the two groups into close proximity, working together on common problems, and demonstrating the value of fire prevention efforts (e.g., reduced false alarms, speedy investigations), have helped in some places to create a sense of teamwork despite different job functions.

## SUMMARY OF STAFFING OPTIONS

The variety of staffing options is seemingly endless. Tasks range from inspection activities to public education or fire investigations. Local decision makers may routinely conduct surveys of "best practices" among other jurisdictions to get ideas about how to most effectively staff and fund their prevention programs. Sometimes, the best ideas come from a jurisdiction that is not comparable in size or population; therefore, those conducting these surveys should be cautioned about being overly restrictive.

As another cautionary note, local decision makers should understand that not every idea can work effectively in all jurisdictions. Local politics, the strength of professional unions, and the needs of their own community are all relevant factors when deciding on the most appropriate mix of service and staffing for their own area. Decision makers should be aware that expecting too much of one individual can result in a lack of expertise or attention that could create legal liabilities and ultimately doom prevention efforts. Expecting one person to adequately conduct a plan review of new construction, enforce the fire code, conduct public life safety and fire presentations, and investigate fires is unrealistic. If resources are limited, the type and scope of prevention programs conducted in any community must also be limited.

## USE OF TECHNOLOGY TO IMPROVE EFFICIENCIES

Many departments are starting to use technology to increase the efficiency of their staff. The use of handheld computers to compile databases for fire code compliance inspections is a common practice in some departments. A variety of

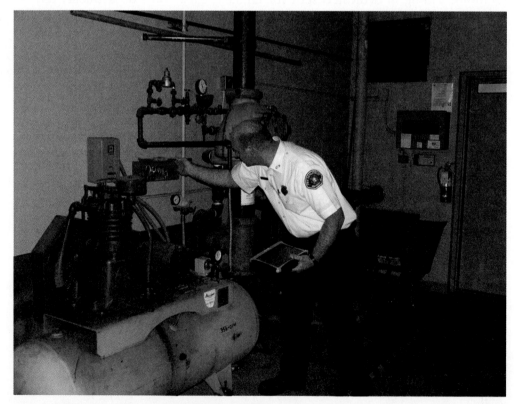

Some jurisdictions are now using handheld computers when performing code compliance inspections. Some model codes are now made available electronically for reference in the field, and inspection forms are also available electronically. (*Vancouver Fire Department, WA*)

hardware and software is available, but finding one solution to meet the needs of *all* fire prevention programs may be difficult. For example, a common database for inspections may be fairly simple to implement. But combining that with a database for new construction, fire investigations, and public education activities may be quite difficult. Still, many have found the use of these products to speed inspections, reduce paperwork, and greatly reduce data entry.

There are a variety of tools that aid in investigation techniques—such as gas detection devices, and even laser devices that can measure the temperature of smoldering debris. Those technologies can be helpful in determining when self-contained breathing apparatus should be worn by fire investigators or emergency responders, by combining harmful gas readings with relative temperatures as an indicator of health hazards.

There are some technologies for public education programs—such as remote-controlled robots, or computers with projection screens and PowerPoint presentation capabilities—that go far beyond the old methods of a standard videotape or DVD. But most often, the best tool we have for prevention programs is the people who conduct them. Their qualifications are probably the most critical factor to consider for organizing and staffing prevention efforts.

## TRAINING OF PREVENTION PERSONNEL

Whether paid or volunteer, personnel cannot perform the functions of their job without proper **training**.

A variety of fire prevention training programs exist at the local, state, regional, and national level for life safety and fire prevention personnel. Courses in code enforcement, fire investigation, plan review, public education, and management of prevention programs are taught at the National Fire Academy in Emmitsburg, Maryland. The International Fire Service Training Association (IFSTA) offers training manuals and associated training packages for each of the prevention disciplines. There are training opportunities through regular conferences at the International Code Council and the National Fire Protection Association—each providing numerous education opportunities to help prevention personnel perform better. NFPA offers a special guide (*NFPA 921*)[2] on how to conduct investigations. NFPA also offers a handbook for public educators, and an inspection manual—as does the International Code Council. The Fire Protection Handbook by NFPA is widely viewed as a source of critical information for fire prevention—and overall fire protection—efforts.

Many states offer more localized training, and in some states, such as Florida, the state association of fire prevention professionals offers both training and **certification** for their members.

*Certification* refers to the process of documenting that training was received, retained, and in some cases put to use. The certification process can involve testing, development of an educational portfolio that is reviewed by other professionals, and in some cases actual field experience. Certifying agencies sometimes provide training as well. The process of certification indicates a higher level of proficiency in the individual receiving it.

The NFPA offers a certification program for inspectors and plan reviewers. The other code-promulgating organizations affiliated with the International Code Council offer certification programs for inspectors who use their codes. The International

**training**
■ educational materials and programs to increase knowledge and skill levels

**certification**
■ documentation that training was received, retained, and in some cases put to use through a process of testing and/or development of an educational portfolio that is reviewed by other professionals

NFPA, ICC, and IFSTA all provide resources to help in training and management of code compliance inspection programs. (*Photo courtesy of ICC*)

Association of Arson Investigators offers a certification program for investigators, as do some other organizations.

Training is considered as a lower level of professional development. Next higher is certification. Some colleges (usually community colleges) offer associate degree programs specifically for fire prevention disciplines. Such a degree can indicate an even higher level of proficiency. Some fire prevention professionals, however, argue that the continuing-education requirements of some certification programs offer more relevancy over a long period of time compared to the one-time demonstration of knowledge required to achieve a degree. Still, each level of professional development can mean something.

That "something" is a professional standard produced by the National Fire Protection Association. A **professional qualification standard** is based on an analysis of the job performance requirements for any particular function in the fire service. NFPA does not produce these standards for every job, but they do exist for fire educators, fire investigators, fire code inspectors, and plan reviewers. There is also one for fire marshals, although this standard is more recent and certifications do not yet exist for it.

Professional qualification standards outline the performance requirements for any given job, including the knowledge, skills, and abilities that a group of professionals have identified as being pertinent to that job. The NFPA manages the process that develops these (commonly referred to as "pro qual") standards. First, industry professionals with an interest in or expertise in a particular job gather and produce a draft of the professional qualification document. After the document is reviewed and voted on by the membership of the organization, it is considered as national standard. Training programs, certifications, and even degree programs are usually built around the foundation of a professional qualification standard.

**professional qualification standards**
■ documents that outline the performance requirements for any given job, including the knowledge, skills, and abilities that a group of professionals have identified as being pertinent to that job

The professional qualification standards for the fire prevention disciplines include:

- NFPA 1031, for fire code inspectors and plan reviewers.
- NFPA 1035, for public educators—currently including those dealing with juvenile firesetting.
- NFPA 1033, for fire investigators. It should be noted that this document is different in scope from NFPA 921, which is a guide to how investigations should be conducted. NFPA 1033 is the professional qualification standard for what an investigator needs to know.
- NFPA 1037 for fire marshals. This standard deals with the fire marshal's job from a management perspective, and with the fire marshal's need to manage all the prevention disciplines (education, enforcement, plan review, and investigations) when so assigned.

Some certification programs do not use the professional qualification standards as a base. For example, the International Code Council currently offers a certification in the fire code that does not adhere to the job performance requirements of NFPA 1031. Thus, it lacks the full scope of duties and skills. The ICC certification tests the ability of the person obtaining the certification to identify the portions of the code that would apply under certain circumstances. It does not measure the ability to apply that knowledge in the field. NFPA's fire inspector certification does adhere to NFPA 1031, and therefore addresses the full scope of duties a fire inspector would face. It requires actual field inspection time as part of the certification process.

Various states offer certifications of their own—though not all adhere to the full professional qualification standards. There are national organizations that serve as an overseer of the quality of certification programs. The International Fire Service Accreditation Congress is one; another is the Pro Board which represents the fire service professional qualification system under their auspices. These two organizations help to ensure that those who undertake certification programs are qualified to do so, and that their methods will withstand professional scrutiny.

The **accreditation** process documents that certification and training programs meet professional standards for quality—in essence by offering an independent review of the qualities and qualifications of those offering certifications and/or training.

accreditation
- documentation that certification and training programs meet professional standards for quality

Local decision makers should realize that each specialty area within a comprehensive prevention program has its own educational requirements, and a need for specialized training. Even volunteers, or emergency response personnel, used for these functions will require training. Requiring too much knowledge (multiple jobs) for any individual means that skill and knowledge levels for some tasks may be insufficient. We need to remain cognizant of the limitations of what one person can effectively manage relative to the complex jobs within the prevention field.

# Funding Options

Funding options for prevention programs are limited. Most are provided for by some governmental agency with taxing authority, which can include local property, income, or sales taxes. In such localities, all of the prevention programs are dependent on the political will of decision makers and of the community they serve. However, prevention programs can sometimes supplement their programs' funding—or pay for them entirely—by any of three other methods.

## GRANTS

Many small departments rely on donations to keep their prevention efforts alive. Even some larger ones supplement their programs with some type of fundraising effort. Grants are one way to obtain funding. Organizations offering grants are usually to some type of nonprofit organization that exists to benefit the community. Some have a local scope, such as the Community Foundation in Vancouver, Washington. Some have national interests, such as the Gates Foundation. Some are directly or indirectly attached to their parent corporations, such as Lowe's Companies, Inc., which provides substantial contributions annually for home safety.

The key to obtaining grants is to understand that there must be a relationship between what a local fire prevention program needs and the interests of the funding organization. For example, the Gates Foundation has a primary interest in education that is predominantly focused on formal educational institutions and oriented toward improving education efforts. This interest has not (as yet) applied to fire safety education because of the foundation's current focus. In other words, if your particular need does not match the funding organization's goals, then even the best-prepared grant application will fail.

There are also grants currently available through the Department of Homeland Security specifically earmarked for prevention programs. A portion of the Assistance to Firefighters Grant Program is devoted to fire prevention programs, but there are usually specific criteria that can increase the likelihood of obtaining funds. Just like the private foundations, those who award AFG prevention grants want the money to do the most good—so they usually target high-risk audiences and prevention programs that have been proven to work on those groups.

Local professionals seeking to obtain grants would be wise to seek professional help. Sometimes that can be obtained from a local college or university, or some civic organization. However, there are some online reference materials such as those produced by SeaCoast that may help those without additional resources to organize a grant application so that it looks professional, increasing the chances of funding. Some organizations, such as the Federal Emergency Management Agency and the AFG grants, have specific grant-writing formats that must be followed.

At the end of the process, there is still someone who must read the grant application and understand it—including the need for what is to be funded and how the funds will be used. And this person (or committee) must be able to tell that the program matches the granting organization's goals for giving and/or community service.

## FUNDRAISING

An almost endless variety of other types of fundraising is available. Many small fire departments still have pancake feeds to raise money for their operations, or their prevention programs. The Muscular Dystrophy Association has developed a positive relationship with firefighters throughout the United States, and they can be seen annually passing around fire boots for local citizens to donate money to help their association. Other departments have used this method to obtain funding for fire prevention efforts. Many have found success in obtaining materials instead of funds. Often people like to see (and feel) the tangible results of their donations, so they purchase or provide funding for specific items such as smoke alarms or public education coloring books.

Sometimes corporate sponsors will provide funding for educational materials if they can place their logo on the materials to obtain some community good will and advertising. In these cases, caution is warranted to avoid a conflict of interest. In one case, a construction company wanted to produce an "after the fire" brochure for a local fire department—with the obvious intent of promoting its construction service to those in need. Other construction companies found that tactic distasteful at best and objected, so the project was halted. However, many corporate sponsors have provided materials with a business logo unrelated to the topic—such as an ice cream company providing the same brochure for how residents can recover after a fire hits their home. In this case, there is no direct link between the target audience and the usual or potential customers of the ice cream company.

In any case, those who are raising funding with dunk tanks and pancake feeds, or by soliciting from companies or individuals in their community, need to be aware of the potential for undue influence or a conflict of interest and seek legal advice if such a question arises. If it even appears to the community that the local fire marshal is being given funding to look the other way on a code violation, damage is done even if that is absolutely not the case. Thus, caution and professional advice are both warranted.

## FEES

Many prevention programs around the nation have moved toward charging fees for their prevention-related services. Most often, that has been in the plan review disciplines. Those involved in the development community have traditionally complained the least about being charged for their services because they can see the direct benefit. They cannot usually obtain an occupancy permit until they receive the plan review and inspection service, and so the fee, if any, is ultimately figured into the cost of construction. However, there is always scrutiny of the level of fees for this or any service, so it is not exactly a foregone conclusion that a fee structure will be adopted by a local jurisdiction.

Fire code compliance inspections are often funded in part by fees as well. These inspection fees are based on the premise that business occupancies receive a higher level of fire protection, because usually private homes are not within the scope of a fire code and cannot legally be inspected. The fee structures can be complex, but there is usually some relationship between the amount of work involved in the inspection and the fee charged: A large building would have a larger fee for inspections than a small convenience store.

Some departments have found a way to charge fees for specific educational programs. Most common are those for first aid or cardiopulmonary resuscitation. The public is generally used to being charged for these types of programs. Some departments also charge for fire extinguisher training (like the Northwest Regional Training Center in southwest Washington) as a way to recoup part of the costs associated with providing that training to the general public.

**Section 16.04.280 Permits and fees**

a. Whenever any permit is required by the fire code, such permit shall be in addition to all other permits or licenses required by law or other ordinance.

b. Permit fees for permits required under IFC Section 105.7 shall be established in VMC Chapter 17.08, Fees Table V Fire Fees, and VMC 20.180.080, Fire Fees.

c. The owner or occupant of buildings that have any of the existing occupancy types listed in this section shall pay a periodic inspection fee, according to fee schedule listed in VMC 16.04.280(c)(1) – (2). For the purposes of this section, "periodic inspection" means an inspection of the existing occupancy types listed in this section, according to the fire code official's pre-set inspection schedule. A "periodic inspection" under this section is not related to any inspection associated with a construction permit, required under VMC Chapter 17.08. For the purposes of this section, "special inspection" means any inspection of the existing occupancy types listed in this section to ensure compliance with newly adopted rules or regulations, compliance with a manufacturer's recall, or any inspection related to a fire code enforcement investigation. There shall be no special inspection fee if a fire code complaint does not result in identifying a fire code violation.

| 1. Scheduled Code-Compliance Inspection, Including First Re-Inspection. | | Occupancy Group 1. B, M & R (Not Including R-3 Occupancies) | Occupancy Group 2. A, E & LC | Occupancy Group 3. F, H, I & S | Special Inspections |
|---|---|---|---|---|---|
| A | 0–3,000 sq. ft. | $60 | $80 | $100 | $50 |
| B | 3001–5,000 sq. ft. | $95 | $115 | $135 | $60 |
| C | 5,001–7,500 sq. ft. | $125 | $165 | $215 | $70 |
| D | 7,501–10,000 sq. ft. | $135 | $205 | $300 | $80 |
| E | 10,001–12,500 sq. ft. | $150 | $235 | $320 | $90 |
| F | 12,501– | $170 | $275 | $335 | $100 |

There are a variety of ways to fund fire prevention programs, usually through tax dollars of some type. But grants, fundraising, and fees are other areas that local practitioners may consider to either fund or supplement their prevention efforts.

# Fire Prevention and Community Efforts

In obtaining community support for prevention programs, there are no solid answers and no one way that works best. But there are some methods that have worked elsewhere—even if their success is not guaranteed.

As with writing grants or fundraising, there must be a connection between fire prevention efforts and the needs of the community. If the community does not see the need for fire prevention, then no amount of asking will help. The tools discussed in Chapters 7 and 12, for planning and evaluating fire prevention programs, are critical for connecting fire prevention efforts to community needs. For example, selecting fire prevention programs that are designed to deal with a particular community problem—identified through research on local loss statistics—will make it easier to justify the community's need for that prevention effort. Gathering data to demonstrate the results of that particular effort will help to ensure that community support is maintained and that local tax dollars are being spent wisely.

Relationships are critical, so developing relationships with decision makers—without upsetting the hierarchy of the fire department—is a positive step to consider. One way to do so is by developing relationships through partnerships in prevention efforts. As has been pointed out, the same target audiences exist whether we are talking about the fire problem, the drug problem, or the crime problem. People at the lower end of the socioeconomic scale account for more injuries, deaths, crime, drug use, and general community service demand than anyone else. If we can find ways to partner our efforts for that community, we can establish relationships with influential people and organizations that may value our prevention programs more than our own departments might.

Cutting a prevention program can be more difficult if it is tied to a large collaborative effort with several community organizations who value it as part of a larger plan. It can be a difficult approach for local jurisdictions to accept, because individual activities are easier to coordinate. Adding partners can be time consuming and messy—because our best method of reaching a target audience might directly and with our single prevention message. But reaching target audiences as part of a larger and more collaborative effort may elicit the community support we need in order to maintain our programs when budgets get tight.

In short, there is a dual relationship between the community's needs and the collaborative relationships we develop to support both the delivery and the maintenance of prevention programs.

# The Importance of Leadership for Fire Prevention Efforts

The field of prevention is a complex area of study and practice. Finding the right leadership for these efforts should never stop at the commonly held belief that a good firefighter—or a good manager—should be able to handle fire prevention program delivery or management. That would be akin to saying that anyone could be a good firefighter off the street—or a good paramedic because they were a good firefighter. The skill set and knowledge requirements are very different.

It should be inexcusable for any department to put their "dead wood" into their prevention programs to finish out their years of employment. The "dead wood" manager will garner no respect from their own organization, let alone the community. A good manager must be competent. Placing someone in the field of prevention without the proper training makes being a good manager virtually impossible. Being a good leader is yet another skill set. Managing the technical aspects of a prevention program is possible with a solid background in the technical side of the job. Mistakes in dealing with people may be forgiven—even by employees—if the technical skills are competent. But a good manager knows how to deal effectively with people—not just the technical aspects of the job. And a leader can take it to a higher level—providing vision and support and unleashing the potential of the people involved in prevention to achieve the most with their resources.

A good leader can obtain the support of his or her own organization—including the emergency responders, by demonstrating the value of prevention programs for their own efforts. In such cases, reducing false alarms or providing quality investigation services that never keep emergency responders tied up waiting too long are ways to demonstrate an effective prevention program and ensure support from other members of the department.

A good leader can garner the support of the community for fire prevention efforts—to help obtain funding, support of materials, and influence to protect prevention efforts when budget cuts are considered.

There are no easy ways to select a good leader, but a beginning would be to select those who are competent and have at least worked in prevention for some time, even if they have not actually achieved some type of professional certification for their field. That, combined with the good qualities of any leader, will help to ensure that prevention programs are professional and respected inside the department and out—and have a chance of survival during tough economic times.

## BUY-IN OF DEPARTMENT AND ELECTED LEADERS

It is worth noting that the best leadership for prevention efforts will be ineffective if those efforts are not backed by departmental or elected decision makers. Effective leaders do in fact attempt to obtain that backing. Without it, funding and support for fire prevention efforts will languish and may even die. There are some anecdotal examples across the nation of fire prevention programs being eliminated entirely because they did not have the support of department and community leaders. One such case occurred in Marion County, Oregon—ostensibly because of budget cuts. And although that cut was made in part to make a point for elected leaders, it nonetheless threw the prevention efforts temporarily into disarray.

## Summary

Staffing, training, and funding considerations for fire prevention efforts are interrelated according to how those efforts are organized. Staffing may consist of paid personnel, either emergency responders working part time or fully dedicated prevention specialists. It may be made up of volunteers, either inside or outside the fire service. In any case, staff should be qualified for the programs they are conducting—whether that is plan review, fire code enforcement, public education, or fire investigations. Training is therefore a critical issue regardless of where the staff is obtained. Training at one level, and certification at another, both should be based on reasonable professional qualification standards that outline the job performance requirements: the knowledge, skills, and abilities needed by those doing the job.

Funding fire prevention programs is often difficult given tight economic times and is usually limited to a few options. Whether via tax dollars, fees, or donations, the money must be available if programs are to have a reasonable chance to succeed. Fire prevention programs must fit within community needs and may be blended with other community efforts in order to develop the relationships necessary to maintain programs when tight budgets force cuts. And good leaders are as important for fire prevention as they are for any other function in the fire service. Good leaders are the best argument against the myth that prevention efforts are largely the purview of the sick, lame, or lazy who could not make it elsewhere in the fire service.

## Case Study

Christmas Valley is a medium-sized community in central California with a population of about 160,000 people. The community has some industrial base, but it is largely commercial and residential occupancies, because it is a "bedroom" suburb for a larger jurisdiction. Recently, the city council in Christmas Valley decided to sever their relationship with the larger jurisdiction and to provide fire prevention services on their own. The larger jurisdiction will continue to provide emergency response, but the city council felt that fire prevention services were lacking and have withheld a portion of the funding to provide services on their own. Beginning with nothing, they are soliciting ideas about how best to construct a fire prevention program and what types of services

will be needed. There are many business occupancies in their jurisdiction that have never been inspected. The city typically refers plan review of new construction to the building department in the county. But now they are creating a building department of their own as well and are considering the best configuration of building code and fire prevention services to meet the needs of their citizens and the development community.

The larger jurisdiction is not happy about the arrangement—and though they have provided company inspections in the past, they recently had to discontinue that service because of the high call volume for emergency responders and the training requirements associated with code compliance inspections.

# Case Study Review Questions

1. Describe the advantages to using paid and volunteer staff for life safety and fire prevention programs, and the problems that might occur between newly hired staff in Christmas Valley and the neighboring jurisdiction, where code compliance inspections were previously done by emergency responders.
2. What qualifications might be necessary for these staff members—and how would you go about finding out about state or national training requirements or certifications?
3. How would you go about developing a training program for life safety and fire prevention staff? Whose support would be necessary, and how would it be funded?
4. Describe which of the prevention disciplines would be important to provide (i.e., plan review of new construction, fire code compliance inspections, public education, and fire investigations) and why.
5. Describe options for funding fire prevention programs, and the pros and cons of each.
6. Describe the pros and cons of a company inspection program.
7. Describe the importance of leadership for fire prevention programs.
8. Describe how prevention fits into the community.
9. Describe how volunteers can be used to support prevention programs.
10. Research and describe some technologies that actually provide for public protection or help to prevent fire problems from occurring in the first place and how they do so.

PEARSON
## myfirekit™

For additional review and practice tests, visit **www.bradybooks.com** and click on MyBradyKit to access book-specific resources for this text! Your instructor may also assign Additional Project work related to topics in this chapter.

Register your access code from the front of our book by going to **www.bradybooks.com** and selecting the mykit links. If the code has already been scratched off, go to **www.bradybooks.com** and follow the MyBradyKit link from there.

# Endnotes

1. *Fire Code Inspections and Fire Prevention: What Methods Lead to Success?* A study on fire code inspections published by the National Fire Protection Association in 1979.

2. *NFPA 921—Guide for Fire and Explosion Investigations;* Quincy, MA: National Fire Protection Association.

# 14

# Public Policy Issues for Fire Prevention Programs

## OBJECTIVES

After reading this chapter you should be able to:

- Identify key components and the definition of public policy relative to fire prevention programs.
- Describe how different approaches to influencing public policy regarding fire prevention.
- Describe a policy "window" and how to take advantage of it.

PEARSON
## myfirekit™

For additional review and practice tests, visit **www.bradybooks.com** and click on MyBradyKit to access book-specific resources for this text!

Dictionary.com defines *public policy* as "the fundamental policy on which laws rest, especially not yet enunciated in specific rules." It further describes *law* as "the principle that injury to the public good or public order constitutes a basis for setting aside, or denying effect to, acts or transactions." To act, or not to act, is the fundamental question that public policy discussions intend to address. And if our government—which is the predominant entity we (our society) put in place to deal with public policy issues—decides to act, then we must address what action to take, how much of it to do, and where and when it will take place. In fact, public policy theory could be applied in any setting where a broad number of people would be affected by decision making that has an impact on societal rules, laws, or norms of behavior. But for the purposes of this chapter, we are limiting the examination to some of the aspects that affect decision making in the governmental setting. And, in this context, we will be examining the aspect of how need drives that decision-making process.

## The Need for Public Policy

In some societies, decisions that affect every member of the public are made by one person. In our society, the way in which decisions that affect the public are made is influenced by a variety of factors, and such decisions are typically not made solely by individuals—though considerable power may be granted to individuals to carry out decision making within certain guidelines. In their purest form, those guidelines are called *public policy*.

For example, the entire set of laws that judges apply to make decisions about who is punished, and how much, are the result of public policy discussions and ultimately a collective (usually legislative) decision about what that public policy will be. For example, lawmakers decide the form of a particular law relating to drug possession, sale, or use. They decide which drugs are illegal (not excepting alcohol) and create laws that society must follow. But how do they arrive at that decision? And what influences are behind the decision-making process?

Specific segments of society who have religious beliefs on the subject of drugs may organize to influence those decisions. People with family members who have suffered from the ravages of drug use may have compelling stories to tell about the negative impacts of drugs, and ideas of what should be done about them. There may be police interests, or prosecutors with a view of the legal system with their own opinions about the scope of the problem and what to do about it. And there is an industry that makes money off of the drugs (legal in this case) that will have money for lobbying efforts. In a very real sense, all of these interest groups would be consulted on, or even drive, decision making regarding drugs.

We need not look very far to see examples of this decision-making process play out in our cultural and societal beliefs and values, and in our laws. Not too long ago, smoking cigarettes on an airplane was considered acceptable. Confined to the back of an airplane, smoking was allowed, even though the smoky air circulated throughout the entire airplane cabin. It is an extreme simplification, but the story of how that behavior and the laws relating to it changed is a microcosm of public policy discussions and decision making.

The people with an active interest in not being subjected to cigarette smoke actively pursued rules that would prohibit smoking on airplanes. Over time they were joined by safety advocates who did not like to see cigarettes on airplanes, where the fire hazard could be a threat to all on board. Health advocates also took a stand with regard to secondhand smoke and became influential advocates for banning smoking on airplanes: A passenger who did not smoke could be harmed by one who did, through secondhand smoke exposure.

In the end, smoking was banned on airplanes, and in fact it is in the process of being banned in many indoor locations throughout the nation for the same reasons. The public policy discussion in this case is about public health and what rules should be in place to protect it. It is a perceived or real need that drives the decision-making process. The behind-the-scenes influence on that process is the point of public policy discussions in this context.

In a societal sense, even an absolute monarch would likely be guided in his or her decision-making process by other people, data, and political forces. In a democracy, the process of influencing decisions is part of the foundation of its very being. Ultimately, a public vote is a powerful tool for establishing public policy, but even a vote is influenced by a variety of forces. Consequently in our society, public policy debates are routine and the public decision-making arena becomes a significant factor in determining how prevention programs are developed and managed.

# Public Policy Formation Models

There are too many resources dealing with public policy to examine in one chapter or even one book. One source that examines the topic is *Understanding Public Policy*, by Thomas R. Dye, published by Pearson Education. That text provides a variety of examples of public policy models that shed some light on the topic, though only a few are listed here, and only briefly so that readers can develop a more thorough understanding of the issue. The goal of this text is to briefly explain some of these theories and tie them into the context of fire prevention program development and management.

**Elite theory** deals with a specific aspect of public policy and how it is influenced. According to Dr. Paul M. Johnson of Auburn University, public policy is really driven by an elite few leaders from key sectors of society. He stipulates that there are interlocking relationships among those from common schools, corporations, or foundations and educational or cultural establishments.[1] In this model, public policy is driven from the top, and when the populace in a democracy has an opinion on a given subject, it is largely influenced by this powerful elite.

**Group theory**, according to Dye, stipulates that interaction among various groups is the central tenet of politics, and therefore of public policy decision making.

In this theory, public health interests and tobacco growers constitute two distinct groups who organized to influence the public policy debate on smoking. Public policy would ultimately be decided by the interaction between them, and a moving balancing point where public policy could change depending on the relative influence of one group at a particular time.

Every day we can see this push and pull in public policy debates between Republican and Democratic parties on a variety of issues—notably health care or foreign policy in our current times. An examination of this aspect of public policy

**elite theory**
- a simple public policy model that stipulates that an aristocracy of a relatively small group of powerful people holds nearly all political power

**group theory**
- the theory that no one group dominates public policy development

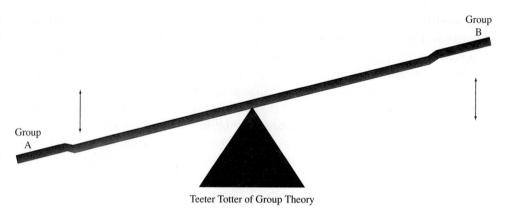

Group theory for public policy debates stipulates that decisions are influenced by competing interests, and that the level of influence they exert over policy decisions may wax and wane over time—creating a constant shift in the way in which public policy decisions are arrived at.

Teeter Totter of Group Theory

theory would yield a study of the influential subgroups that in turn influence the larger group (e.g., the Republican or Democratic parties).

This seesaw effect can be seen in our own public policy debates. In prevention terms, the groups with an interest in protecting the environment may be on one side of the urban wildfire hazard issue—and those advocating for defensible space may be on the other. Our own push and pull or seesaw of public policy discussion can play out in our prevention arena. And the same is true of interests opposed to residential fire sprinklers as a mandatory requirement in all new construction of one- and two-family dwellings, and the fire service and industry interests advocating for those same mandatory requirements.

**institutionalism**
■ the theory that governmental institutions have a predominant level of influence in determining public policy

**Institutionalism** refers to the structures, duties, and functions of specific governmental institutions and how they influence public policy. According to Dye, a policy does not really become *public* policy until it is adopted, implemented, and enforced by some governmental institution. Those with a group-theory view of public policy debates would argue that the rules changing smoking activities in the United States were derived not from institutions, but from other societal forces with an interest in (and a passion for) that particular topic. The same could be said of laws related to drunken driving. Mothers Against Drunk Driving (MADD) is not a government institution. But when private organizations become large enough, they may become a political force of their own.

In addition, government institutions are often involved in public policy discussions before decisions are made. It is important for fire prevention managers and fire service leaders to understand the institutionalism in public policy debates, and

The institutional view of public policy stipulates that governmental institutions are both an influencer of public policy debates and the mechanism used to carry them out once decisions are made and adopted.

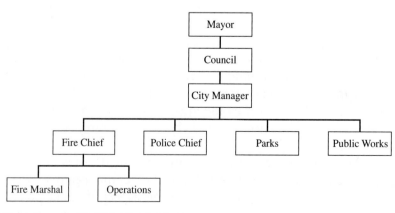

the role of the fire service within those debates. The fire service may exist to enforce public policy, but it also has a significant role in developing it.

These institutional models can play out in a variety of settings for the fire prevention field. Assuming that the institution (meaning the fire department) has sole authority to determine public policy would be incredibly naïve. Fire prevention professionals who have tried mandating rules controlling the possession and use of consumer fireworks independent of a public policy discussion have discovered that just because the fire department is the institution with a distinct interest in the topic, that does not empower it to make autocratic decisions that affect the public.

In the public policy arena, we would be wise to understand how these theories play out in the real world. There are many examples: developing and adopting a fire or building code; mandating rules that apply to fire protection contractor professional qualifications; mandating fire sprinklers; outlawing smoking; or developing rules for creating defensible spaces around homes to protect them from wildfires. These are all fire prevention issues that will be affected in one way or another by the public policy debate that ensues when the issues arise—and when we are advocating our positions on them.

# Fire Prevention Aspects of Public Policy

Readers have already been exposed to one of the most significant influences on public policy where fire prevention programs are concerned. The **focusing event** can galvanize general public sentiment to drive public policy decisions.

An example is the Beverly Hills Supper Club fire that occurred in May 1977 in Southgate, Kentucky. In that fire, the third worst nightclub fire in U.S. history, 165 people died. The tragedy galvanized public sentiment and pressure, which led then-Governor Julian Carroll to make serious changes in code enforcement because of the results of the investigation: the building had not been built according to safety standards, and the owners had allegedly avoided regulation. It also led to a change in the electrical code regarding aluminum wiring at the National Fire Protection Association—which produces the National Electrical Code. That private institution was also influenced by this disaster—a focusing event that helped shape public policy for public and private entities alike.

**Partnerships** are another aspect of public policy that affect fire prevention and fire service administrators.

Partnerships in this context are another way of looking at elite theory—recognizing that there may be people in the community who have much greater influence on public policy than the fire service does. For example, who in the community would likely have an interest in creating cigarettes that are less likely to cause fires? The answer has been found in legislative efforts begun in New York state under then–State Fire Marshal Jim Burns and carried on by the National Fire Protection Association. The coalition formed to support legislation for "fire-safe" cigarettes developed over many years, through the efforts of such people as Andrew McGuire, the director of the Trauma Foundation at San Francisco General Hospital. Andrew and others had been urging the tobacco industry for years to create a type of cigarette that would go out when it was not being puffed on by a smoker. Some studies had provided evidence that these cigarettes would be less

**focusing event**
- a major event such as a fire loss that captures public and political attention, generating pressure to change public policy

**partnerships**
- different groups that come together to influence public policy, rather than competing to determine what public policy will be

likely to cause fires. Smoking-related fires were then, and still are, one of the principal causes of fire deaths.

The coalition that the NFPA formed included representatives from the American Association of Retired Persons, the American Burn Association, the American College of Emergency Physicians, the American Fire Sprinkler Association, the Asian American Hotel Owners Association, and many others with a common interest in reducing fire deaths. Some of the members from the medical community and public health advocates were, and are, more interested in stopping smoking. But the common ground also gave them an interest in stopping smoking-related fire deaths—hence their natural inclusion as partners in a coalition that ultimately picked up far greater political support than the fire service would have been able to achieve alone.

The fire-safe cigarette coalition (founded by the National Fire Protection Association) estimates that about 99 percent of the U.S. population is now protected by laws that require these fire-safe cigarettes, which are constructed so that they will go out when they are not being smoked. No one is saying that these laws will eliminate fire deaths from cigarettes entirely. In fact, currently many states are facing an influx of unregulated cigarettes that are cheaper and do not have the built-in safety features that make them go out when they are not being smoked. But the tremendous progress made on the subject since the issue first arose in the 1970s was finally made possible not by one focusing event, but rather by many that occurred over time and the successful partnerships that came together to affect this aspect of public policy.[2]

Of note are the social forces at work—including a very bad economy, which is driving people to make their own cigarettes rather than buying those that are "fire safe." Some jurisdictions are beginning to see an influx of cigarettes made by Native American tribes with claims of sovereignty that supersede the regulations governing the production of "fire safe" cigarettes.

political support

■ *direct* political support for fire prevention policies by elected officials with the authority to mandate public policy

**Political support** is of course driven by many factors that have already been mentioned, including those that are central to the elite, group, and institutional theories of public policy. But it is important enough to mention specifically.

Powerful friends may influence elected officials, as may groups and partnerships that come together to do so. But fire service leaders should be mindful of the fact that they are part of the institutionalized government run by politicians, and are therefore a part of the direct political influence that can gain or lose support from elected officials. However the relationships are developed, formally or informally, the direct relationship between fire service and political leaders is critical in public policy debates. And trust is the foundation of those relationships.

Political leaders cannot always do the homework necessary to fully understand every issue that comes before them. Many are diligent students of public policy. Only the foolish are content to remain ignorant of issues, voting out of friendship or some arrangement of favoritism. As one politician (former City Councilperson Gretchen Kafoury of Portland, Oregon) described it, the art of gaining political support from other elected officials is another aspect of public policy influence. But the relationship between the fire service leader and the political leadership is one of trust. Sooner or later they must trust that the fire service leader knows what he or she is talking about and will keep them informed of issues that could change the basis of that relationship.

Gretchen was the city council member responsible for the Fire Bureau in Portland in the 1990s. She pushed through a vote of council members on a public safety bond to retrofit fire stations so that they would be able to withstand an earthquake. Her relationship with the fire service leaders was predicated—in her own words—on being kept informed. "No surprises" was the motto of her term in office. If an issue facing the fire department could end up on the news, she wanted to know about it first. Her ability to head off negative publicity—or at least to deal with it proactively—was what helped to keep her in public office. She was the type of politician who operated not out of a sense of public importance, but rather public duty. And her reliance on personal relationships with fire service leaders—founded on trust—was a great case study in how the direct political support of elected officials can profoundly affect the operations of a local fire department.

Gretchen Kafoury later went on to establish public policies for requiring and funding fire sprinklers in some affordable housing projects, low-cost housing being her number one political priority. She did so because she trusted the officials providing her with information on the topic. The message was carried by her senior staff members—notably Dr. Jim Marshall—and sold on the premise that doing so was necessary because Portland could not afford to let fires destroy valuable housing stock for the low-income segment of society.

This aspect of developing personal relationships with political leaders—and gaining direct political support—should not be overlooked by fire prevention leaders. It is based largely on time, trust, and communication. But fire service leaders have a head start on that credibility-based relationship because they were hired to do the job. Unless they are subject to a specific political agenda (such as a politician out to reduce fire service expenses), there may be a natural relationship with elected officials, if only because the fire service is a fellow part of the institution of government.

Fire service leaders are often called on to testify to the merits of public policy decisions, and as such, they should not overlook the importance of relationships with elected officials as decision makers.

Relationships are critical, but the credibility and trust between fire service leaders and their elected officials, or the general public, is also founded on the **public need** for those services.

R. Wayne Powell, formerly of the U.S. Fire Administration and now working for Marriott International—said it this way: "No data—no problem; no problem—no money." It was his way of saying that unless we were able to demonstrate the actual need for public funding, no amount of relationship building would overcome other funding priorities.

Demonstrating the need for prevention efforts comes mostly from the data we collect. Much more is said about this topic in Chapter 12 on evaluating fire prevention programs, and in Chapter 7, planning for fire prevention programs. The data we collect demonstrating the scope of the fire problem is going to be helpful in demonstrating the need to do something about it. Other ways to demonstrate need are potentially powerful. Real life, personal and anecdotal examples can sometimes be the most compelling when convincing others of the need for prevention programs.

One such example is the group referred to as *Common Voices*. This group of family members who have been burned, or who have lost loved ones to fire, was formed to be the voice of victims and advocates for improved fire prevention practices. They are focused especially on fire sprinklers as a solution to the nation's fire problem, but they have also been effective advocates of other prevention methodologies. Their personal stories can stir an emotional response where no amount of data will sway public attention or opinion.

Ultimately, fire prevention leaders will need to recommend specific programs and budgets for public approval. Figuring out how to demonstrate the need for these programs—especially during difficult economic times—is an important element of their ability to influence public policy relative to prevention programs.

Common Voices provides personally compelling anecdotal examples of the importance for fire prevention efforts—especially fire sprinklers—based on their own tragic experiences. (*Courtesy of Common Voices*)

# Summary

Public policy issues related to fire prevention are the forces behind the decision-making process as to which programs get funded, and how much they receive. There are varying theories of influence. Elite theory subscribes to the concept that only a powerful few truly influence public decisions and laws, whereas group theory holds that it is a combination of groups and competing forces that drive public policy. The institutionalist theory of public policy recognizes that there are government (and private) institutions that are a key factor in influencing public policy decisions, as well as in carrying them out. But all the theories have practical implications for fire prevention leaders.

When we are deciding which of the comprehensive list of fire prevention programs to provide—and how to obtain the resources to provide them—we are entering the world of political influence and public policy. Choosing to fund regular fire code compliance inspections, for example, over public fire and life safety education programs is a policy decision that might be driven by our ability to support only one approach rather than the other. When making decisions of that nature, individual fire prevention leaders could potentially be making their own public policy decisions, but they are in turn driven by other forces, such as an economic downturn that forces cutbacks in all prevention disciplines.

But to obtain additional resources to keep programs afloat, or to advocate for new programs such as fire sprinkler requirements, fire service leaders must understand the forces influencing public policy—and in turn influence that policy to support their prevention efforts. Understanding the power of a focusing event, such as a local fire disaster, or the fact that influential partners in the community might help sway public decision makers is fundamental to a public policy strategy for influence. Recognizing that direct political support could be obtained through personal relationships, and that everyone needs to understand the need for prevention programs, is also a base for influencing public policy decisions. Clearly articulating the need for prevention programs is the foundation on which all other influencing factors will be built. If this need is not understood, no personal relationships, partnerships, powerful friends, or focusing events will sway public decision makers to support fire prevention programs.

For prevention programs, the public policy debate is about decisions: which programs to have and fund. Those decisions are based on need—and on how much risk any community is willing to accept. In the end, the plan review, fire code enforcement, public education, and fire investigation programs that are disciplines of a comprehensive approach to prevention are all about managing community risk. The ultimate goal is to reduce losses, but there are so many factors involved that it might be impossible to demonstrate those kinds of results. So risk management becomes the basis for how the prevention programs operate, and local decisions must be made about what level of resources to put into which disciplines.

Those decisions will be heavily influenced by the amount of funding that the community has available to put toward their fire protection efforts. In that sense, prevention programs will compete for funding with the emergency response needs for fires, emergency medical incidents, hazardous materials incidents, and the wide array of other emergencies a modern fire department must face. And as the push for results grows, caused by limited funding, prevention practitioners and fire service leaders will find themselves in a never-ending cycle of demonstrating need, maximizing their influential abilities, figuring out what works best—and advocating for the most efficient and effective prevention programs possible.

# Case Study

## Public Policy

For years, the fire service in the United States has been advocating for residential fire sprinklers. As a mechanism for reducing fire losses, fire sprinklers (when combined with working smoke alarms) have been the most demonstrably effective means of reducing fire deaths and fire damage in the United States. These facts are supported by studies conducted by the National Fire Protection Association over years of analyzing national fire loss data. A study by the National Bureau of Standards (now the National Institute of Standards and Technology) in the United States found that the estimated likelihood of dying in a fire is reduced by 82 percent when both a smoke alarm and residential sprinkler are added to a home that had neither.[3]

But a crucial factor has been elusive for the fire service in all these years, since fire sprinklers were first being advocated in the late 1800s. The expense of the systems has always generated opposition to their installation, and especially to laws that mandate their use.

In recent years, residential fire sprinklers have become more affordable thanks to advances in technology, and the work of technical committees (usually at the National Fire Protection Association which houses the installation standards for fire sprinkler systems). The debates among technical committee members and different industry representatives are about which types of systems should be allowed. When combined with research and field data that demonstrates what works and what does not— the back and forth exchange in these committees have over time produced changes in those standards that allow more inexpensive systems to be produced for residential settings.

But whether or not they should be required has been a public policy debate in local communities and at the national level for many years. Some communities, such as Scottsdale, Arizona, have had residential fire sprinkler requirements for many years. Sprinklers are considered one of the basics of their fire protection services, based on the concept that in-home fire sprinklers will function more quickly than any fire department response and ultimately save the taxpayers money in long-term costs for fire protection. The same concept was applied in Vancouver, British Columbia, in the 1990s. Then–Fire Chief Don Pamplin helped push through a fire sprinkler requirement for all new construction in 1990. He advocated for that ordinance on the basis that it would help contain long-term costs for fire protection for the community. He was heavily criticized by local fire union officials for this stance, as well as by the building industry, which did not want to see a mandate that would increase their costs. According to an article published by Sean Tracy, the regional manager in Canada for the National Fire Protection Association,[4] the enactment of the fire sprinkler ordinance in 1990 has led to a significant reduction in fire deaths. He estimates that they may be as much as 50 to 66 percent below what they were before the ordinance. Equally important, the City of Vancouver did not suffer the economic problems opponents of fire sprinklers predicted.

Even with these local examples, mandating fire sprinklers in residential occupancies in the United States was also being done mostly at the local level, despite fire service and industry efforts to get them required by code. However, in 2005, with little fanfare, the building and fire codes promulgated by the National Fire Protection Association were all changed to include requirements for fire sprinklers in all one- and two-family dwellings. This process of policy influence also had been in the making for many years and included a variety of partners. But it was not widely heralded because at the time the NFPA codes were not widely adopted nationally.

The principal target of those who wanted to see these requirements in the model national building and fire codes remained the International Code Council and the codes they promulgated. In previous years, the International Fire Code had been changed to include a requirement for residential fire sprinklers. However, the construction requirements for one- and two-family dwellings were held in the International Residential Code. Consequently, the long public policy debate at the national level began to focus attention on residential fire sprinkler code requirements and settled on the historic vote at the ICC code hearings of 2008 in Minneapolis, MN. There were actually two ICC hearings that year, but according to the International Residential Code Fire Sprinkler Coalition (www.ircfiresprinkler.org), the final debate occurred in Minneapolis and centered on RB 64-07/08 and the International Residential Code. The coalition that worked for this requirement was successful at obtaining a two thirds majority vote of all in attendance at those hearings, and the model code included requirements for fire sprinklers in all one- and two-family dwellings in the 2009 edition of the IRC, with an effective date of January 1, 2011.

The IRC Fire Sprinkler Coalition, led by President Ron Coleman and Executive Director Jeff Shapiro, was instrumental in organizing the effort to get enough voting members to the hearings in Minneapolis and to arrange for cohesive and compelling testimony to convince anyone who had not already formed an opinion on the matter to vote in favor of the code requirement. The coalition was not alone. There were numerous individuals and other organizations behind the requirement and code change. Among them were the National Fire Protection Association, the International Fire Chiefs Association, the International Association of Firefighters, and the Home Fire Sprinkler Coalition (www.homefiresprinkler.org)—an organization that existed to promote voluntary installation of fire sprinkler systems. Not all were part of the actual debate at the code hearings, but much of the work about the value of fire sprinklers was predicated on the work that these and other organizations had already done regarding the benefits of fire sprinklers.

Notably, the fire sprinkler industry (the National Fire Sprinkler Association and the American Fire Sprinkler Association) helped provide funding to the IRC Fire Sprinkler Coalition, thereby causing a subdebate about how the model codes would be influenced by industry interests.

The code requirement passed—but the public policy debate about requiring residential fire sprinklers was far from over. As it currently stands, the National Association of Home Builders and some other development-related interests continue to fight the requirements on several levels. According to the IRC Fire Sprinkler Coalition press release on the topic, this debate is taking place state by state.

## ATTENTION FIRE MARSHALS—Homebuilder tactic against residential sprinklers

It has come to our attention that homebuilders are beginning a move to prevent adoption of the 2009 IRC provisions for residential fire sprinklers by introducing state legislation that would block new local adoptions of these provisions. Such bills were recently filed in Arizona (HB2267) and North Dakota (SB2354). If your state fire service organization utilizes the services of a legislative monitor, please alert them to be on the lookout for anti-residential sprinkler legislation, and if you come across something, please let us know.

That item was placed on the IRC Web site in January 2009. Since then, many states have faced either legislative requirements or challenges to the adoption of the 2009 IRC by varying code councils in each state. Some have been successful—based in part on the argument that the cost of fire sprinklers raises the price of housing in a market currently undergoing a crisis of monumental proportions.

More recently, the ICC hearings in October 2009 dealt with efforts on the part of homebuilders and others to remove the code provisions that require residential fire sprinklers. The code panel itself voted to uphold the requirements in the body of the code—which was a significant change from previous code development cycles. Some of the panel members changed their previous votes to be in favor of fire sprinkler requirements, stating the will of the collective membership. When opponents of the requirements called for a floor vote, the overwhelming response (due to very large numbers of fire department participants) was in favor of keeping the requirements in place. The struggle over the requirements, however, continues at the state level in legislatures, and in code-adopting policy groups all across the United States.

Despite the tremendous solidarity of all the stakeholders in the effort to require fire sprinklers, the debate will likely continue for some

There are many local ordinances requiring residential fire sprinklers. Some that have been in place for some time are under attack by development interests who say that fire sprinklers are too expensive for the benefit they provide. *(Courtesy of Anne Arundel County Fire Department)*

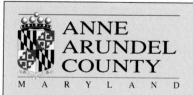

**ANNE ARUNDEL COUNTY**
M A R Y L A N D

**Fire Department**
8501 Veterans Highway
Millersville, MD 21108
Phone (410) 222-8200   Fax (410) 987-2904

**John R. Leopold, County Executive**
John Robert Ray, Fire Chief

## **NEWS RELEASE**

**For Immediate Release:**
January 6, 2009

**Contact:** Lieutenant Shawn Jones

### LANDMARK LEGISLATION SAVES LIVES IN ANNE ARUNDEL COUNTY
Mandatory Residential Sprinklers in New Homes Wins Approval

Millersville (MD) . . . County Executive John R. Leopold has won a major victory for the citizens of Anne Arundel County as the County Council approved a measure to require residential sprinklers in all new homes built in Anne Arundel County beginning later this year. "This is a tremendous moment in our County. I commend Councilman Vitale for her sponsorship of this important public safety initiative, as well as the other five councilmen who voted in favor of the legislation. Passage of this law places a firefighter in every new home in our County who will remain on duty at all times. There can be no greater priority than saving lives," the County Executive said in learning of the vote.

Fire Chief John Robert Ray was equally thrilled. "This legislation will save lives. While smoke alarms provide the early warning needed to escape from a fire, residential sprinklers protect the lives of those who cannot escape including the infant in a crib, the child who does not wake up to the sound of the alarm, the elderly and the infirmed. I would like to thank County Executive Leopold for his steadfast support of this bill and his willingness to allow us to advocate for its passage. This is a great day in our County's history that will last for generations to come."

The bill, first introduced in late 2008, was supported by every major fire service organization on a federal, state and local level. The bi-partisan 6-1 vote reflects that this is not a political issue or one based on economics but instead is focused squarely on public safety. Members of the Anne Arundel County Volunteer Firefighters Association, Professional Firefighters IAFF Local #1563, Maryland State Fireman's Association, Maryland State Fire Chief's Association, the Maryland State Fire Marshal and a host of emergency responders were present in Council chambers when the vote was taken.

Residential sprinklers have been identified as the single most effective way to reduce civilian and firefighter fatalities in structure fires. Designed to allow occupants time to escape, these devices frequently control or even extinguish a fire prior to the arrival of responders. Every year, over 3000 Americans die in residential structure fires—more than were killed on 9-11. Fires account for billions in direct property loss annually—more than was caused by Hurricane Katrina. A recent National Fire Protection Association study found the cost of installing residential sprinklers in Anne Arundel County to be the equivalent of installing upgraded carpet or countertops—approximately $1.61 per square foot.

*An All Hazards Response Organization, Committed to Your Safety* years to come. But local ordinances are still being passed to require sprinklers—such as in Anne Arundel County in Maryland, where several counties have now passed residential fire sprinkler requirements.

Virtually all proponents of fire sprinklers would agree that no matter how this particular issue turns out, the debate will benefit from the public's understanding of the value of residential fire sprinklers.

## Case Study Review Questions

1. Describe the advantages to using partners to help influence public policy debates on residential fire sprinklers; fire code adoption and aggressive compliance programs; public education programs designed to improve the installation of smoke alarms; and effective fire investigation efforts.
2. How would a debate on the importance of prevention programs be shaped by an analysis of the fire loss data for a given jurisdiction?
3. How would you go about developing a local ordinance requiring fire sprinklers in all residential occupancies? How would you go about developing a local ordinance requiring fire sprinklers in all residential occupancies as Vancouver, BC did in 1990?
4. Assuming budget cuts, describe which of the prevention disciplines would be most important for a community to provide

(i.e., plan review of new construction, fire code compliance inspections, public education, and fire investigations) and why.
5. Describe the elements of elite theory regarding public policy.
6. Describe the elements of group theory regarding public policy.
7. Describe how fire prevention programs may help influence public policy.
8. Describe how the rise in no-smoking laws was driven by public policy forces—and how they succeeded in their efforts to make smoking in public places illegal.
9. Describe a "focusing event" and its impact on public policy.
10. Given the case study, describe the status of adoption and implementation of residential fire sprinkler requirements in your own locale or state.

PEARSON
## myfirekit™

For additional review and practice tests, visit **www.bradybooks.com** and click on MyBradyKit to access book-specific resources for this text! Your instructor may also assign Additional Project work related to topics in this chapter.

Register your access code from the front of our book by going to **www.bradybooks.com** and selecting the mykit links. If the code has already been scratched off, go to **www.bradybooks.com** and follow the MyBradyKit link from there.

## Endnotes

1. Johnson, Paul M. (n.d.). A Glossary of Political Ecomony Terms. Retrieved from http://www.auburn.edu/~johnspm/gloss/elite_theory.

2. N.A. (n.d.). "The Coalition for Fire Safe Cigarettes." Retrieved from http://www.firesafecigarettes.org/categoryList.asp?categoryID=9&URL=Home%

20-%20The%20Coalition%20for%20Fire%20Safe%20Cigarettes.

3. Ruegg, Rosalie T., and Sieglinde K. Fuller (November 1984). "A Benefit-Cost Model of Residential Fire Sprinkler Systems," NBS Technical Note 1203. Gaithersburg, MD: U.S. Department of Commerce, National Bureau of Standards, Table 6.

4. Tracey, Sean (n.d.). "NFPA Impact: The fire service in Canada needs to get behind residential fire sprinklers." Retrieved from http://www.firefightingincanada.com/content/view/1338/213/.

# GLOSSARY

## A

**Acceptance inspections**—physical visits done in the field to ensure that what is approved on the plans is actually built in the field

**Accreditation**—documentation that certification and training programs meet professional standards for quality

**Administrative duties of the fire marshal**—the management aspects of the job, rather than the technical expertise of any given portion of it

**Alternate materials and methods**—that portion of the code that allows local officials to accept different ways of achieving the same overall level of safety provided by the code

**Anecdotal data**—personal examples, informal witness reports, news articles, etc., that are more limited, but often more revealing because they can represent one particular case in human terms that captures attention of decision makers

**Appeal process**—outside board of citizens within the authority having jurisdiction

**Applied research**—research used to solve a particular problem

**Arson**—the crime committed when an individual has deliberately set a fire

## B

**Balloon construction**—the type of construction made of framed wood—without fire stops to prevent a rapid spread of fire—that was evident in older housing stock

**Benchmarking**—the process of comparative analysis—one jurisdiction to another or one to itself

## C

**Cause determination**—the process of identifying how fires begin

**Certification**—documentation that training was received, retained, and in some cases put to use through a process of testing and/or development of an educational portfolio that is reviewed by other professionals

**Coalition development**—the process of bringing partners outside the fire service together to support fire prevention programs and goals

**Coalitions and partnerships**—groups or organizations coming together to provide more resources than one partner could muster alone

**Code administration**—managing the code enforcement process

**Code development**—process of creating code provisions

**Code enforcement**—the process in which local jurisdictions have the authority to order compliance with fire and/or building code requirements

**Combined prevention programs**—the partnering of various techniques, such as education and enforcement—or technology and education—for best results

**Company inspections**—when on-duty emergency response personnel conduct fire code compliance inspections

**Construction plan review process**—activities associated with making certain that new construction or remodels comply with appropriate building and fire codes

**Culture**—the beliefs, norms of behavior, and traditions that form the social glue of a people

**Cultural anthropology**—the science that deals with the origins, physical and cultural development, biological characteristics, and social customs and beliefs of humankind

## D

**Direct fire losses**—the loss of fire in dollars

**Diversity and fire prevention**—taking different languages and culture into account, and reaching out to others to create diverse partnerships that aid fire prevention efforts

## E

**Elite theory**—a simple public policy model that stipulates that an aristocracy of a relatively small group of powerful people holds nearly all political power

**Environmental analysis**—review of a variety of elements that constitute the "environment" of a governmental entity such as a fire department

**Evaluation**—the gathering and analysis of data (statistical or anecdotal) that provides evidence of a program's impact or outcome

**Evidence-based decision making**—the process of gathering information and analyzing it before making decisions about fire prevention program design

## F

**Fire-department influences**—a significant factor because of the level of importance they place on prevention programs

**Fire protection contactors**—private sector contractors who inspect, test, and maintain fire protection systems

**Focusing event**—a major event such as a fire loss that captures public and political attention, generating pressure to change public policy

**Formative evaluation**—the process of gathering information about the scope of the fire problem before design begins, including its root causes and the details about the people who are affected by it

## G

**Generational culture**—the differences in values, beliefs, and norms of behavior that exist between differing age groups

**Government Accounting Standards Board**—a private nonprofit organization devoted to developing and promoting model performance measures that aid local decision makers in documenting the results of programs supported by tax dollars

**Governmental influences**—regulatory power that governments exert over fire prevention efforts

**Group theory**—the theory that no one group dominates public policy development

## H

**Hierarchy of needs**—the theory produced by Abraham Maslow in 1943 that describes a tiered pyramid of needs that all humans have

**Historical planning efforts**—national planning processes that have led to improvements in fire prevention in the United States throughout our history

## I

**Impact evaluation**—the mechanism of measuring changes that the program is intended to produce in the target population

**Indirect fire losses**—those that can be extrapolated from the direct costs associated with replacement

**Inspection process**—procedures followed by code enforcement personnel to conduct an inspection

**Institutionalism**—governmental institutions have a predominant level of influence in determining public policy

**Intangible fire losses**—those we suffer when we cannot describe the loss in financial terms

**Integrated risk management**—planning for a combination of risks and a balance of methodologies (e.g., emergency response and prevention) to deal with them

**Investigation**—activity by which cause and origin is determined; generally refers to fires and explosions

## J

**Juvenile firesetters**—younger children who most often set fires out of a sense of curiosity without an understanding of the consequences, but occasionally do so because of some emotional disturbance

## M

**Master planning**—overall strategic and tactical planning for fire department operations

**Multicultural society**—one that is made up of numerous cultures—all maintaining some sense of cultural identity—while melding certain aspects of their cultures into a common set of traditions and beliefs

## O

**Occupancies**—distinct business operations. A building might contain one or many occupancies, each with a different operation and sometimes different occupancy classifications

**Operational permits**—required to outline storage and use requirements for specific hazards identified in the code

**Organizational culture**—unwritten values and norms of behavior for an organization that are the foundation of its operation

**Outcome evaluation**—the mechanism of determining how well a program achieves its ultimate goals

## P

**Partnerships**—different groups that come together to influence public policy, rather than competing to determine what public policy will be

**Performance-based design**—design based on goals for fire and life safety rather than relying on the prescribed requirements found in the model codes

**Performance measurement**—documenting results programs produce in specific terms

**Plan review**—checking construction plans for compliance with applicable codes, and conducting acceptance inspections in the field to ensure that construction practices match what is shown on the plans

**Political support**—*direct* political support for fire prevention policies by elected officials with the authority to mandate public policy

**Potential risks**—those that do not have a high incident rate, but would represent a significant risk (e.g., hospitals) should a fire occur

**Preservation of evidence**—collection and protection of accurate notes and physical evidence

**Private-sector influences**—lack regulatory authority, but still influence fire prevention efforts in a variety of ways

**Process evaluation**—the mechanism of testing whether a program is reaching its goals

**Professional qualification standards**—documents that outline the performance requirements for any given job, including the knowledge, skills, and abilities that a group of professionals have identified as being pertinent to that job

**Public education**—results-oriented programs that raise the cognitive awareness levels and ultimately change behaviors of targeted audiences

**Public information and public relations**—programs that help build positive relationships and generally improve the image of the fire service

**Public need**—the real or perceived need from a broad spectrum of the public, which heavily influences public policy direction

**Pure research**—involves fact finding and verification of theories for its own sake

## R

**Real risks**—those pointed out by a statistical history of events

**Regional planning**—a collaborative process of planning for mutual benefit with those who may provide us resources in time of need, or who may need help themselves

## S

**School-based educational programs**—those programs that are general—covering several safety topics—and age appropriate. They may involve using a specialized curriculum where the teachers instruct their students, or presentations done in schools by fire safety professionals

**Seasonal educational programs**—programs designed for topical and regularly anticipated fire problems such as Christmas tree fires

**Significant fires**—large fires with multiple deaths

**Standard of cover**—a document, produced through a process similar to master planning, that arrives at local decisions for deployment of emergency response personnel and resources

**State influences**—provide leadership on fire prevention programs outside the code arena

**Statistical data**—the historical bits of data that provide information about patterns of fire cause

**Steps in public education planning**—*Identification* refers to identifying problems we need to address through research; *selection* refers to selecting achievable and specific goals; *design* refers to designing programs toward those goals and specified audiences; *implementation* refers to development of an implementation plan; and *evaluation* refers to development of an evaluation plan. All five steps are completed before implementation actually begins

**Strategic planning**—broad directions in a traditionally identified process that includes stakeholder influences on the plan; different from tactical planning

**T**

**Tactical planning**—more detailed than strategic planning; provides a "how to" approach once the broad directions are identified

**Targeted educational programs**—programs designed for specific messages and specific audiences

**Testing laboratory listing**—a stamp that indicates a particular product has met the performance requirements and safety standards for that product

**Training**—educational materials and programs to increase knowledge and skill levels

**W**

**Workload, efficiency, and effectiveness measures**—the specific terms government performance auditors usually use when describing the results of programs